Applied Sampling

This is a volume of

Quantitative Studies in Social Relations

Consulting Editor: Peter H. Rossi, University of Massachusetts, Amherst, Massachusetts

A complete list of titles in this series appears at the end of this volume.

Applied Sampling

Seymour Sudman

Departments of Business Administration
and Sociology and the Survey Research Laboratory
University of Illinois, Urbana–Champaign
Urbana, Illinois

ACADEMIC PRESS, INC.
Harcourt Brace Jovanovich, Publishers
San Diego New York Berkeley Boston
London Sydney Tokyo Toronto

ACADEMIC PRESS, INC.
1250 Sixth Avenue, San Diego, California 92101

United Kingdom Edition published by
ACADEMIC PRESS, INC. (LONDON) LTD.
24/28 Oval Road, London NW1 7DX

Library of Congress Cataloging in Publication Data

Sudman, Seymour.
Applied sampling.

(Quantitative studies in social relations)
Bibliography: p.
Includes index.
1. Social surveys. 2. Social science research–United States. 3. Sampling (Statistics) I. Title.
HN29.S688 309′.07′23 75-34460
ISBN 0–12–675750–X

PRINTED IN THE UNITED STATES OF AMERICA

87 88 89 9 8 7 6 5

Contents

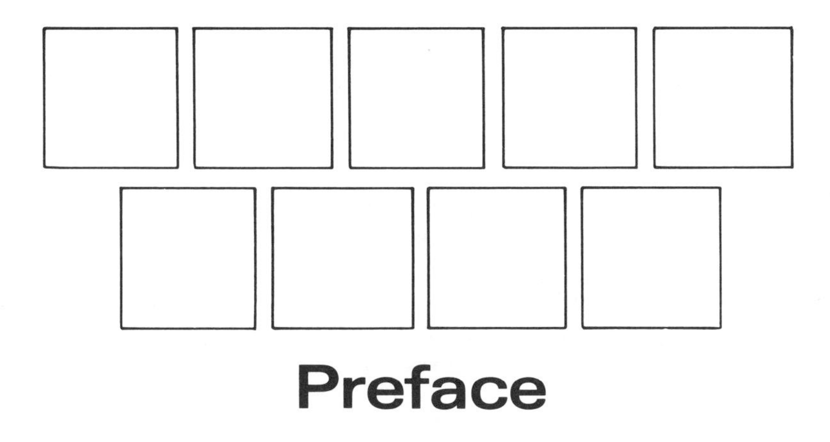

Preface

This book is intended for researchers who have limited resources and statistical backgrounds and who wish to maximize the usefulness of the data they obtain. My impression is that in many cases these researchers are aware that powerful sampling procedures are available, but believe that they are unable to use them because they are too difficult or costly. Instead, incredibly sloppy ad hoc procedures are used, often with disastrous results. Even the excellent book by Kish (40), full as it is of useful and practical advice, assumes a scale of operational quality beyond the scope of most of these researchers.

This book emphasizes procedures that are much less expensive and only slightly inferior to the most precise methods; or, from the other perspective, methods that are substantially better than the ad hoc methods and only a little more costly. The sampling procedures described here are not intended for use by large governmental agencies for major policy decisions. There, as we shall see in the first chapter, very large and precise samples are required because of the magnitude of the decisions depending on the results. For these users, the classic books by Hansen, Hurwitz, and Madow (35) and by Cochran (14) serve admirably.

This is not intended to be the major text for a course in sampling, nor as a reference book for sampling experts. It will be useful to students and researchers who are actually engaged in planning a study. For this reason, the book omits mathematical derivations and concentrates on examples. It is assumed, however, that the reader has had an introduction to statistics and is familiar with means and ratios, standard deviations, correlation coefficients, and, most importantly, can follow the ever-present summation notations of statistics without much difficulty.

The material in this book has evolved from a series of workshops and individual consultations over the past decade both at the National Opinion Research Center, University of Chicago and at the Survey Research Laboratory, University of Illinois. Many of the examples come from participants in these workshops as well as colleagues. I am grateful for their help.

I am also glad to acknowledge my intellectual debts to the classic sampling books by Cochran; Hansen, Hurwitz, and Madow; and Kish that have already been mentioned. These works are indispensible to any serious sampler or student of sampling. Although all three books are well-organized and clearly written, they do demand a higher level of statistical background than is assumed in this book.

The Bayesian ideas about the value of information are derived from the work of Robert Schlaifer, William Ericson, and from my former professors, Harry Roberts and Leonard Savage.

I am also grateful for the careful reading and comments on earlier drafts of my colleagues Edward Blair, Robert Ferber, Ron Czaja, and Jeffrey Goldberg of the Survey Research Laboratory and Martin Frankel, Norman Bradburn, and William Kruskal of the University of Chicago. Finally, I am especially grateful to Peter Rossi for his help and encouragement both in developing the framework for this book and suggestions for improving the specifics.

Naturally, I am solely responsible for any errors or ambiguities that remain in the text, and I would appreciate hearing about any errors discovered. I do not apologize, however, for trying to write carefully about some relatively sloppy sampling procedures. When probability sampling methods were first becoming popular about three decades ago, the pendulum swung heavily in the direction of precision and away from cost concerns that had characterized the cheap but sloppy quota sampling procedures. Since we are more sophisticated now, it is possible to consider methods for increasing the cost-effectiveness of sampling without major sacrifices in quality.

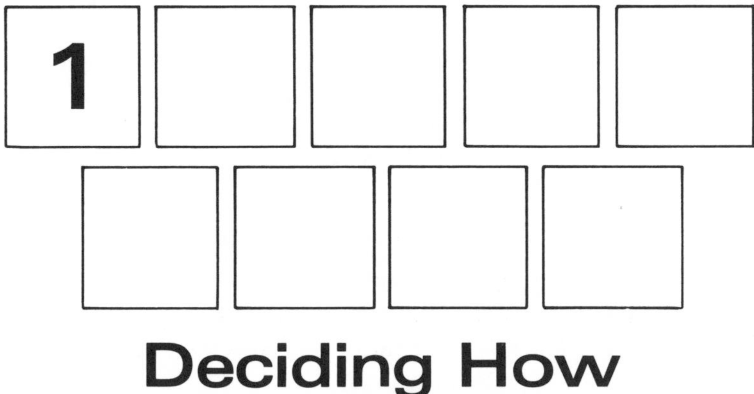

Deciding How to Do the Study

1.1 INTRODUCTION

This book is intended for researchers who wish to use survey methods for data collection, but who have limited experience and resources. In far too many cases, these researchers are aware that powerful sampling methods are available, but believe they cannot use them because these methods are too difficult or expensive. Instead, incredibly sloppy ad hoc procedures are invented, often with disastrous results.

Since the 1950s procedures have been developed that are much less expensive and only slightly inferior to the most precise methods; and from another perspective, they are substantially better than ad hoc methods and only a little more costly. In this book, we concentrate on these alternative methods.

The general procedure used in this volume is to introduce a sampling idea and then to illustrate it with real-world examples. In this chapter, we start with 10 examples of reasonable sample designs for various purposes and stages of research, and contrast these with some examples of inappropriate samples. This is followed by a discussion of the earliest decisions that must be made in designing a study—defining the population, deciding what field methods, and if a new or existing field organization should be used. The chapter ends with a description of sampling for special populations or for those in limited geographic areas.

Most readers will be faced with hard choices of the best procedure to use with very limited resources. Chapter 2 concentrates on these small, special samples and illustrates that even here there are methods for improving quality.

Chapter 3 discusses the simplest sampling procedures primarily involving the use of available lists, and methods for correcting flaws in lists. Chapter 4 introduces the notion of reducing costs of data gathering through geographic clustering, but notes that clustering reduces the reliability of results. Chapter 5 gives both theoretical and practical answers to the first question generally asked in planning a study–how big should the sample be? The theoretical answers to this seemingly simple question are surprisingly difficult, even in the simplified versions given here.

Chapter 6 deals with stratification procedures for improving the efficiency of sampling, that is, reducing sampling variance when costs, variances, or prior information differ by subpopulations called *strata.* Chapter 7 discusses multi-stage samples of the general population, such as those normally used in public opinion research. While it is unlikely that readers of this book will ever need to design a national sample, this chapter describes procedures for using census data for selecting a state or local sample. Chapter 8 gives procedures for computing sampling errors from all kinds of samples. Since the simple formulas given in beginning statistics courses do not apply to complex samples, the procedures discussed involve the ideas of replication and pseudoreplication that depend on the basic definition of sampling variance.

The final chapter, "Special Topics in Sampling," describes four types of sampling that do not fit in the other chapters but are useful in a wide variety of situations–probability sampling with quotas, screening for rare populations, snowball sampling, and panel sampling. At the end of each chapter, suggestions are given for additional readings. After the first two chapters, the remaining chapters are relatively independent of each other, so the reader who wishes to skip some chapters or revise the reading order is free to do so.

1.2 HOW GOOD MUST THE SAMPLE BE?

It should be made clear immediately that there is no uniform standard of quality that must always be reached by every sample. The quality of the sample depends entirely on the stage of the research and how the information will be used. At one extreme, there is exploratory data gathering used in the process of generating hypotheses for later study. At the other extreme are large-scale continuing studies used to supply the input for major policy decisions of the federal government. Obviously, the levels of accuracy required differ in these two extreme cases. Thus, one of the earliest decisions that must be made in planning a study is how good the sample must be.

To illustrate the differing levels of accuracy required, we first present a series of examples ranging from high to low levels of accuracy, where the quality of the sample seems to be appropriate to the requirements. In contrast, we then cite several examples of samples where the quality is either too high or too low.

Example 1.1 The Current Population Survey

The largest continuing personal sample of households in the United States is the Current Population Survey (CPS) conducted monthly by the Bureau of the Census (89, 91, 92). The sample is located in 449 sample areas comprising 863 counties and independent cities. Some 60,000 housing units are designated for the sample each month; about 52,500 of them, containing about 105,000 persons aged 16 years and older, are occupied by households. The remainder are vacant or converted to nonresidential use.

The very large and carefully controlled sample is necessary because the data from the CPS are the only source of monthly estimates of total employment and unemployment, as well as the only comprehensive source of information on the personal characteristics of the total population between the decennial censuses (86, 88).

Major government economic and welfare programs are influenced by changes of a few tenths of a percent in CPS data from month to month. For this reason, the sample is large enough so that the sampling errors of the total estimates of unemployment are only about 0.1%.

The sample has grown from 25,000 total units a month in 1943 to 40,000 units in 1957 to 60,000 units in 1967 because of the increased concern about local area economic and social problems. There is some probability that even larger samples will be taken in the near future instead of the more frequent censuses that also have been proposed. Although the current budget for the CPS is about $15 million annually, there is general agreement that this level is required.

It is interesting to note that, in addition to being cheaper and more timely, the data from the CPS generally are considered more accurate for many purposes than is the decennial census. This is because continuing attention is given to the hiring and training of an experienced staff of interviewers and to the careful design of detailed questionnaires so that responses obtained are less subject to error. It is not possible, of course, to provide the same detail about small geographic areas from any sample, no matter how large, as from a complete census. A discussion of general population samples like the CPS is given in Chapter 7.

Example 1.2 Equality of Educational Opportunity (The Coleman Report) (15)

The Coleman Report was the result of Section 402 of the Civil Rights Act of 1964, which required the U.S. Commissioner of Education to conduct a survey and report to the President and the Congress concerning the lack of availability of equal educational opportunities for individuals by reason of race, color, region, or national origin in public educational institutions at all levels.

The survey, which involved some of the leading social scientists in the United States, was perceived from the beginning as having major policy implications for enhancing the educational opportunities of minorities. For this reason, every effort was made to design a high-quality sample.

The sample for this survey was located in 398 counties and metropolitan areas, 1170 high schools, and the elementary and junior high schools feeding into the selected high schools. Within the selected schools, all pupils in Grades 1, 3, 6, 9, and 12 were sampled, using self-administered questionnaires. The total initial sample size was approximately 900,000 pupils, half white and half nonwhite. Black students were oversampled relative to their proportion in the total population, since the aim was to compare black and white students (see Chapter 6). The final sample, excluding refusals from schools and individuals within schools, was about 645,000.

Major problems arose during the survey because of the refusal of some large city school systems to cooperate, causing some concern about sample biases. Nevertheless, this study continues to be used extensively by social scientists and policymakers.

Example 1.3 Supplemental Studies for the National Advisory Commission on Civil Disorders (The Kerner Commission) (82)

During the middle 1960s, riots in urban black ghettos became increasingly common. The Kerner Commission was appointed to investigate these disorders and report on their causes and possible solutions. Much of the data gathering involved use of newspaper files and informal interview with community leaders. Two large and carefully planned surveys financed by grants from the Ford Foundation were part of the supplemental studies prepared for the Kerner Commission. In contrast to the previous example, however, it was recognized that the roots of the riots were spread through the entire society and that it was less likely that specific policy decisions would result from the surveys. Thus, the sampling was less precise than the Coleman school sample, although still of high quality.

Both surveys in this example were conducted in 15 major American cities: Baltimore, Boston, Chicago, Cincinnati, Cleveland, Detroit, Gary, Milwaukee, Newark, New York (Brooklyn only), Philadelphia, Pittsburgh, San Francisco, St. Louis, and Washington, D.C. These cities were selected judgmentally as being those most crucial for an understanding of the riots. They ranged widely in the severity of riots during the previous year. There were, however, no Southern or Southwestern cities, and Los Angeles was omitted, as were important parts of some metropolitan areas, such as Oakland, California, and the rest of New York City. Although the sampling within cities was done very carefully, the limitations of the study result from the initial selection of cities.

The population of black and white residents was studied by Campbell and Schuman of the Survey Research Center, University of Michigan (12). A total of 5759 interviews were completed in the 15 cities, 2582 with whites and 2814 with blacks. An additional 363 interviews with whites came from the suburbs of Cleveland and Detroit. Respondents were located using area sampling procedures (discussed in Chapter 7) to identify a household and one respondent between the ages of 16 and 69 from that household. Since the goal was to compare black and white respondents, black households were oversampled.

In a concurrent study of the same 15 cities, Rossi and his associates (69) studied the attitudes of about 2250 people who provide services in the ghetto and hire ghetto residents. In each city, the desired sample was 150 respondents—40 from the police, 30 merchants and 30 major employers, 20 social workers and 20 teachers, and 10 political workers.

The decisions regarding whom to include in this sample and what sampling rates to use with the various groups were to some extent arbitrary but primarily reflected ghetto residents' concerns. Thus, the highest samples were taken from the police because concerns about police treatment received widespread mention as a cause of riots. The police, teachers, and social workers were sampled from lists provided by the police department, school system, and public welfare department in each city. In four of the cities, there was substantial difficulty with the police sample, but all the other groups cooperated. Employers were chosen from a list of the 100 largest employers in the area, obtained from Dun and Bradstreet. Retailers and political party workers were selected by using rather loose quota procedures.

Example 1.4 The Education of Catholic Americans (34)

Greeley and Rossi directed a national study of Catholics in 1964, financed by grants from the Carnegie Corporation and the U.S. Office of Education, and conducted by the National Opinion Research Center, University of Chicago. The key research issues investigated were: (*1*) Does parochial education make better Catholics of the students who attend? and (*2*) Are there divisive effects of Catholic education that weaken the consensus of American society and lead to a ghetto mentality among Catholics? Although Greeley is a Catholic priest, this study was not funded by Catholic sources, but it was anticipated that the results would have an impact on the debates concerning the role of religious education in America.

The study was done with a national area probability sample of Catholics, aged 23–57, who had been located in a study on adult education conducted the previous year by NORC. The availability of this group as the result of the previous year's screening substantially reduced the costs of locating respondents.

From the initial screening of 12,000 households, a total of 2753 Catholic respondents were located and interviewed. A control sample of one-fourth this size was taken from Protestant respondents in the same geographic areas as Catholics. In addition, 1000 self-administered questionnaires were left at the homes of other Catholic respondents and, in households where there were adolescents, another self-administered questionnaire was left to be filled in by those currently in high school. Finally, a questionnaire was mailed to a randomly selected sample of 1000 readers of *Commonweal* so that the analysis could compare this group, presumably the liberal intelligentsia, with the total population of Catholics.

Example 1.5 The Gallup Poll (28,63)

Probably the best-known of all continuing U.S. polls, the Gallup Poll reports not only on presidential elections but on all major public issues as they arise. The poll is syndicated and appears in major newspapers, each of which pay a small fee for the publication rights. The syndication fees support the service, and determine the size and quality of the sample.

The sample for each survey consists of 1500 adults selected from 320 locations, using area sampling methods. At each location, the interviewer is given a map with an indicated starting point and is required to follow a specified direction. At each occupied dwelling unit, the interviewer must attempt to meet sex quotas. While this block quota sampling procedure is not completely unbiased, since it misses people who are less likely to be home (see Chapter 9), it appears to provide results that are near enough to the true values to give politicians and the public a sense of public attitudes about an issue. The election predictions, when compared to the results, indicate the general accuracy of the sample.

All the remaining examples will describe samples of limited geographic areas. Such samples are examined in more detail in Chapter 2. In Example 1.6, limiting the study to a sample of Illinois is clearly a function of the study's topic. In the later examples, the geographic limitations are necessary because of financial or operational constraints.

Example 1.6 Public Perception of the Illinois Legislature (19)

In this study conducted by the Survey Research Laboratory of the University of Illinois, the sample consisted of 400 Illinois residents of voting age. Although, to achieve equal sample reliability, the sample size for a state or local geographic area would need to be virtually as large as if the study were a national sample of the United States, one generally finds that local samples are smaller. That is, although public attitudes toward a state legislature are an important topic, the level of research funds available is smaller for a state than for a national study.

Since the funds for this study were limited, the interviewing was done by telephone, using statewide Wide Area Telephone Service (WATS) lines and limiting the sample to households with phones. This introduces a sample bias since some 10% of the state's population are without phones. However, given the scope of the study and resources available, the bias is not too large to significantly reduce the usefulness of the results. As we shall see in Chapter 3, the use of telephones is becoming increasingly important where resources are limited and phone availability is high. Note that, while phone samples are appropriate for this example, they would be inappropriate for the Current Population Survey, where much higher accuracy is required. Even for the CPS, however, some phone interviewing is conducted if the respondent cannot be reached by a personal interview.

Example 1.7 The National Labor Relations Board (NLRB) Election Study (30)

Getman and Goldberg, with a grant from the National Science Foundation, studied the effects of pre-election behavior of employers and unions on the voting behavior of employees at NLRB elections to determine if the union should serve as the exclusive bargaining agent. Aside from the general research interest in voting and organizational behavior, the study has important policy implications. If the study finds, as preliminary results indicate, that employees are not particularly attentive to the campaign and are not quick to catch nuances of coercion, the NLRB may revise its rules on what constitutes an unfair labor practice.

Given the importance of the study, one might think that a national sample of elections would be chosen, but the study was limited to Illinois, Indiana, Kentucky, and Missouri. This limitation was necessary because of the logistics of conducting the study. Interviews were conducted both before and after the election, and there was only a short time, frequently less than a week, between selecting an election for study and starting the interviews. Because of the complexity of the issues, the interviews were conducted by law students from Illinois and Indiana closely supervised by the principal investigators. Thus, time and money constraints limited the geographic area of the study.

Even within the geographic constraints, the sample of 35 elections and some 1500 employees was not a random sample of all elections. The size of the bargaining unit and the expected intensity of the election were also taken into account. Size is an objective variable, but the predicted intensity of the election was determined by Goldberg and Getman on the basis of their knowledge of the firms involved as well as preliminary discussions with union and management representatives. It is possible that this could introduce some biases. (The discussion of this example continues in Section 1.4.)

Example 1.8 The Detroit Area Study

A rapidly growing trend in social science departments of major universities is the use of practical experience to teach research methods. The oldest and best-known of these programs is the Detroit Area Study (DAS) operated by the Survey Research Center, University of Michigan. Each year a different project is conducted, with the topic selected from among competing proposals submitted by faculty members. The project is executed in the field by a combination of professional interviewers and beginning graduate students; the students also process the data and analyze some portion of it by mutual agreement with the project director. Frequently, papers prepared to fulfill class requirements are published later in professional journals.

While the University of Michigan provides some of the operating funds for the DAS, additional funds are usually obtained from outside granting agencies. The size of the sample and the complexity of the design thus vary from year to year, depending on the aims of the study and the resources available, but generally a probability sample of around 500 interviews is obtained. Because of the limited time students have available, restricting the study to the Detroit Standard Metropolitan Statistical Area makes it possible for each student to do some interviewing under careful supervision of the Survey Research Center. Obviously, this geographic restriction determines the kinds of projects that are possible as well as limits the generalizability of the findings.

Example 1.9 Unfunded Doctoral Dissertation Research

Instead of a specific example, we describe a composite of the kinds of research samples that are possible assuming only the effort of the doctoral candidate and out-of-pocket expenditures of less than $1000, usually paid by the candidate with some help from his department. If a general population sample and face-to-face interviewing are required, the samples are typically in a single place and the number of respondents usually ranges from 200 to 300. All the interviewing is done by the candidate or by the candidate with a few helpers, either paid or unpaid. The selected place is chosen to be easily reached by the candidate in order to avoid travel costs. There is usually a city directory available to reduce sampling costs.

If the questionnaire lends itself to telephone interviewing, the sample can be spread over a larger area, particularly if a WATS line is available. The sample is selected either from telephone directories or by the use of random digit dialing (see Chapter 3). Since no travel time or costs are incurred, a larger sample of up to 500 cases is possible. The actual sample size depends on the complexity and length of the questionnaire as well as phone charges, if any.

Frequently, special populations, such as professionals (lawyers, dentists, teachers) or organizations (schools, hospitals, business establishments), are chosen for analysis. This generally reduces the attainable sample size since much more effort must be expended to locate the special population and to obtain cooperation. Thus, with face-to-face interviewing, samples of as few as 50–100 cases are frequently all that can be reached. With phone interviewing, samples of 200–300 are possible if WATS lines are available.

With some professional groups, mail surveys are possible, particularly if the survey has the endorsement of the national or local leadership. Depending on the availability of lists, the sample can be national, regional, or local. Sample sizes of 500–1000 are used since the major cost is postage and printing. To obtain a reasonable cooperation rate, at least one and usually two or three follow-up mailings are necessary. Finally, if cooperation still is low, additional phone calls are made to a subsample of about one-third of nonrespondents. (See Chapter 6 for the explanation of why this is an optimum procedure.)

Example 1.10 Pilot Tests, Exploratory Research, Motivational Research

The lowest-quality samples generally consist of 20–50 respondents usually chosen at the convenience of the researcher. If household respondents are used, the interviewer is free to select the household from anywhere in a broad geographic area, although sometimes a block or census tract are specified. Sometimes a church or other voluntary group will be used for either a self-administered questionnaire or a group interview; if the researcher is connected with a university or school, the respondents may be the students in a classroom.

There is little in this book that can help researchers who use these kinds of samples except some indication of when poor-quality samples are appropriate. They are appropriate at the earliest stages of a research design, when one is first attempting to develop hypotheses and the procedures for measuring them. Then, along with reading the literature and discussing ideas with colleagues, friends, and relatives. exploratory data gathering is worthwhile. Any sort of sample may be useful when very little is known. Just a few interviews can pinpoint major problems with questions and dimensions of the topic that the researcher may have ignored.

1.3 INAPPROPRIATE SAMPLE DESIGNS

Whether or not a sample design is appropriate depends on how it is to be used and the resources available. In some cases, it may be fair to say that the sample design is appropriate for the available resources, but that the analysis and generalizations made from the sample go too far.

Consider the student who is doing unfunded research. It would be inappropriate for that student to attempt or be advised to attempt a large national study. The resources available are just not adequate for the task. Any such study attempted is very likely to be badly executed with very low cooperation rates from respondents. Almost all researchers will agree that a small study well-designed and executed is superior to a large study that has been botched. On the other hand, it is generally possible for a student writing a Ph.D. dissertation to do more than the lowest-quality exploratory research discussed in Example 1.10. Frequently, research is labeled as "exploratory" merely to protect it from criticism directed against a poorly designed or executed sample.

At the opposite extreme from students are federal government research and evaluation projects, but here too inappropriate samples are sometimes seen. For example, during the funding of Model Cities programs by the Office of Economic Opportunity, an evaluation of the programs' effectiveness was required. As a standard procedure, OEO required a population sample of 1% of the households in a model city. Although there may have been some political reasons for this decision that were never made clear, there was no statistical rationale for the procedure. This rule resulted in samples that were probably too large in the largest cities in the United States and too small in smaller metropolitan areas. In all these places, sampling variability depends not on the percentage of the population *but almost entirely on the sample size alone* (see Chapter 8). If the agency had wanted the same level of reliability in each model city, then identical sample sizes should have been taken. Even assuming that the larger cities were more important than the smaller ones, decisions about sample size should have been made on the basis of the accuracy required to evaluate the effectiveness of various programs, not on the basis of the population size.

On the other hand, many economists and statisticians feel that the sample now used for the Consumer Price Index (CPI) is both poorly drawn and too small to reflect accurately important price changes in the U.S. economy (85, 87). Unlike the Current Population Survey, which is in 449 sample areas, the CPI is in only 56 cities and metropolitan areas selected judgmentally and to some extent on the basis of political pressure. In reality, these cities represent only themselves although the index is called the Consumer Price Index for Urban Wage Earners and Clerical Workers, and most people believe the index reflects prices everywhere in the United States. Within cities, the selection of retail outlets is done judgmentally so that the possible size of sample biases is unknown.

The greatest use of inappropriate samples, however, is by professors in the social sciences. Given the availability of a large sample of *captive* respondents in beginning classes, many academics never consider the use of broader and more representative samples, even if resources are available. For these professors, a high-quality study is one in which the sample consists of students at several schools selected because the study director has friends there. In many cases, the task required of students is completely inappropriate to their current status, such as asking the student to assume a leadership role in a business organization or to make a household buying decision.

Even where the task is appropriate, the subordinate relationship of the student to the instructor leads to exaggerated responses that cannot be duplicated in real-world samples. For example, the well-known Rosenthal (68) studies, which have been conducted to indicate experimenter effects, have used students as experimenters and subjects. Most of the observed effects probably would vanish if a general population sample were used.

1.4 THE USE OF BIASED SAMPLES FOR SCREENING

In some cases, the use of very small and poorly chosen samples may be justified as the first stage of a screening process if the directions of the biases are known. Consider the research process for discovering new drugs to treat various forms of cancer. Hundreds of such drugs have been proposed and it is impossible to test each drug on a large, carefully drawn sample of patients. Instead, the drugs that appear most promising on the basis of experiments with animals are tried on very small samples of patients at hospitals near the researcher. If the new drug is ineffective on this sample, it will usually be discarded since there are so many other drugs to try and there is little reason to believe that the drug would work on some other group.

If, on the other hand, the drug is effective with some of the patients, the drug would then receive more careful testing from larger samples with careful controls for placebo effects and the selection of patients. In other words, the sample biases in this case are expected to be in the direction of overstating the effectiveness of the drug.

A more cheerful example of the same procedure is the use of employees of a company to test new products that are produced in research and development. If the employees dislike the new product, or even if they express some reservations, the product is in serious trouble and probably will fail to reach the marketplace. The sample biases are such that one would generally expect employees to be more interested and enthusiastic about a company's new product than would be the general public. If the employees are enthusiastic, the product is then tested further on real-world populations.

Finally, let us consider Example 1.7, the NLRB Election Study. If a random selection of elections were made, only a minority of these would be likely to be intensely disputed and result in unfair labor practices. Then, if the study directors found no evidence of an effect on employee voting behavior due to employer and union behavior, critics could claim that this occurred only in intensely disputed campaigns, and that the sample of these campaigns was too small.

Thus, since only 35 elections can be observed, a deliberate decision has been made to oversample highly disputed elections. If there is no evidence of any effects on employees in these elections, it is unlikely that there would be any effects in calmer elections. If, on the other hand, there are some effects on employees, the sample would tend to overstate them. The design optimizes the probabilities of overturning the study directors' prior expectations and is in the broad scientific tradition in which a scientist with a hypothesis tries his hardest to disprove it.

1.5 DEFINING THE POPULATION

A sample is most generally defined as a subset from a larger population. This would suggest that, before thinking about samples, one already has a clear picture of the universe or population (the terms are used interchangeably) from which the sample is to be selected. Unfortunately, researchers frequently forget to make explicit the universe they wish to study, or assume that the universe corresponds to the sample selected. This leads to strange definitions of universes, such as the universe of all college freshmen in beginning psychology classes, or the universe of readers of a specific magazine or newspaper, when in fact the real universe under study is the total adult population of the United States. It is better to have a clear sensible definition of the target universe and then to carefully describe the sample than to have a misshapen universe definition to fit a strange sample.

The first step in defining a population is to decide whether it is a population of individuals, households, institutions, transactions, or whatever. The source from which the data are to be collected need not be identical to the population definition. If a mail questionnaire is sent to college presidents asking for information about riots on the campus, the population is of colleges and not of college presidents. Similarly, one household member may report on household income or savings if the universe is household spending units, or may report on employment of individual members of the household if the universe is individuals. While most of the time the choice between a universe of individuals or households will be clear, there is a fuzzy middle area. Should studies of consumer behavior, media usage, and leisure time activities use populations of individuals or households? The decision is a difficult one and will depend on the specific purposes of the study, but the important thing is that the issue be considered carefully and decided as the study is being planned.

Once the unit of analysis has been determined, the next decisions involve what units to exclude. The following criteria should be considered:

Geography. Unless the study relates to a policy question for a local area, such as a state, city, or other political unit, the geographic definition of the universe is usually the entire United States, although even this limits the generalizability to other countries. Cross-national studies and sampling are so complex, however, that they are beyond the scope of this book.

Age of Individuals. Generally, some minimum age is established. For attitudes on public issues, the minimum age is usually 18; for studies of employment, it is usually 16; for media readership, either 10 or 12. Ordinarily there is no maximum age, but there can be if the study deals with women of childbearing age or with newly married couples.

Other Demographic Variables. Sex, race, marital status, and education are other variables sometimes used to define a population. While sex is seldom ambiguous, almost all other variables must be defined carefully. For example, if the sample is limited to white respondents, are Puerto Ricans, Indians, Filipinos, and other orientals to be included or excluded? If the study is of black respondents, should African and Latin American blacks be included?

Other Individual Variables. Citizenship, voter registration, and intent to vote are crucial variables to define in any studies of election behavior, but may or may not be important for other studies of public opinion.

Household Variables. If the unit of analysis is the household, one must first define a household. The census definition that is usually used defines a household as everyone living in a housing unit, where a housing unit is one in which tenants do not live and eat with any other persons in the structure and in which there is either (*1*) direct access from the outside of the building or through a common hall or (*2*) complete kitchen facilities for the use of the occupants.

Even this very precise definition needs further explanation. Mobile homes, trailers, tents, boats, or railroad cars may contain households, but are excluded if vacant (as are vacant housing units), used only for business, or used only for vacations. Occupied rooms in hotels and motels are included if the residents have no usual place of residence elsewhere.

The characteristics of the household also must be defined carefully. If the population is limited to households of a single race, a decision must be made about mixed households. Sometimes the race of the household is defined as that of the head of the house, while other times, the households are excluded. If the requirement is that the household be intact—that is, with both a husband and a wife—decisions must be made on how to treat couples who are separated, either legally or otherwise. For example, how would one treat a household in which the husband is a sailor on a nuclear submarine and is not home now but lives at home when his ship is in port?

The most difficult family characteristic to define is income, particularly in poverty areas where it is hard to determine who is and is not a permanent household member. Since a definition of poverty depends on both income and size of family, whatever definition is used to determine whether or not an individual is included in the family should also be used in determining if that individual's income should or should not be included.

Even if the sampling unit is institutions, decisions must be made as with households and individuals. With organizations, there is often a minimum size limit, such as business firms with fewer than five employees are excluded from the universe. In sampling a population of colleges and universities, one needs to decide whether or not to include 2-year colleges, religious seminaries, military universities, postgraduate institutions, unaccredited schools, business colleges, and so on. The decision would depend, of course, on the aims of the study.

1.6 PROBLEMS WITH OVERDEFINING THE POPULATION

One should avoid overdefining a population if it is not critical to do so. Beginners sometimes think that setting narrow age and income limits makes the study easier to conduct and reduces variability in the results. While there may then be less variance in the sample data, there is also no possibility of generalizing to a broader universe. Operationally, narrow definitions of the universe greatly increase the cost and difficulty of finding respondents. There are no published lists and no easy ways of finding men between the ages of 20 and 40 with incomes between $8000 and $12000. This requires the screening of a very large sample of the general population and asking questions that are both difficult to ask at the beginning of an interview and always subject to response errors. The simpler the universe definition, the easier and less costly to find the sample.

1.7 OPERATIONAL DEFINITIONS OF THE POPULATION

Although a population should not be defined to mirror a convenient sample, the definition should be possible to implement in the field. Thus, rather than defining a population as consisting of all women still capable of bearing children, it is preferable to define the population of women between the ages of 12 and 50. While this definition by age may exclude a few women who are capable of childbearing and include some who are not, the more general definition is not operational. Similarly, in a study of the effects of noise on residents near an airport, it is better to define the population as all those living within 1 mile of the airport rather than defining the population as all those affected by airplane noises. It should be recognized that statistical inferences from the sample can be made only about the sample population. To the extent that the sample population differs from the target population, inferences about the target population must be subjective.

While it is useful to have census information about a defined population, this is not always possible or necessary. Studies of religious groups, criminals, and homosexuals may have well-defined populations although no census material is available. Samples selected from these populations are usually not representative of the total special population. Most studies of criminals are done in prisons, a practice that biases the sample toward those criminals who are more likely to be arrested and convicted. Studies of homosexuals usually have been conducted in bars or with members of organized groups like the Mattachines, and are biased toward those members of the population who are most active socially. The careful studies done with these groups point out the sample deficiencies rather than attempt to revise the definition of the population.

1.8 WHAT FIELD METHODS SHOULD BE USED

The nature of the research will determine the most efficient way of collecting the data, and this in turn determines what kind of sampling will be optimum. Thus, data collection methods should be considered prior to making sampling decisions. Generally, one considers the advantages and disadvantages of face-to-face interviewing, telephone interviewing, mail questionnaires, or some combined method.

Some studies can only be done with the interviewer and respondent face-to-face. If the interviewer must hand the respondent a list of alternatives or pictures from which to choose, face-to-face interviewing is necessary until visual telephones become common. If the sequence of the questions is important to avoid order effects, self-administered questionnaires cannot be used. For example, on a personal interview, it is easy to determine the salience of problem areas by first asking a general question, such as "Are there any problems with living in this neighborhood?" and recording the free responses. Then one can go on to ask specific questions, such as "Are people around here concerned about air pollution?" On a self-administered questionnaire, the respondent usually looks over the entire form before beginning, so a later mention of air pollution can influence the respondent to mention it on an earlier question.

Self-administered questionnaires also cannot be used when a respondent's knowledge is to be tested or when it is important that a respondent not be in the presence of nor consult with other household members, as in a study of household decision making where one wishes to get independent answers from husbands and wives on how they behave jointly.

If the nature of the study is not a limiting factor, the next dimension to consider is the cost of alternative data-gathering procedures. Mail methods, if possible, are generally cheaper than phone or face-to-face interviews, although not as cheap as the novice might believe. Any sensible mail procedure will include not only the postage and mailing costs for the initial mailing, but the same costs for at least two follow-up mailings to nonrespondents. In each mailing, a copy of the questionnaire should be included; thus mail surveys require that about twice as many questionnaires be printed. Experiments also have shown that the use of stamps rather than reply envelopes increases response, so the full cost of postage should be budgeted for mailing both ways and for multiple mailings.

A precise mail budget requires an estimate of the number of responses to the various mailings; if 60% respond to an initial mailing, the costs of subsequent mailings obviously will be less than if only 30% respond initially. As a rough rule of thumb, if less than about 30% are expected to respond initially, the study should not be done by mail if a careful sample is wanted. Experience suggests that the probability of response for initial nonrespondents on subsequent mailings is about the same as the initial probability. If 50% cooperate on the first mailing, 50% of the remaining group, or 25% of the total, will cooperate on the second mailing, and 50% of the last remaining group, or 12.5% of the total, will cooperate on the third mailing. If one can estimate the initial response to a mail questionnaire, either from past experience with similar questionnaires for similar groups or, even better, from a small pilot test, then a mail budget can be accurately estimated.

One other cost factor must be considered in mail surveys—the time required to complete the survey. Assuming one allows 3 weeks between mailings, more than 2 months will pass before the last responses trickle in. If additional personal follow-ups are then required, an additional 2 months are necessary. If there are staff members with no other tasks who are waiting for the results, then the cost of the delay can be considerable.

The costs of phone procedures involve interviewer time and possible toll charges. The major advantage of phone over face-to-face procedures is the elimination of travel time in locating respondents. This reduces both the cost and time required to collect the data. Supervision costs also are reduced, while the effectiveness of supervision is increased if the phoning is done from a central location. If a call is made and the respondent is not home or is unavailable, not much time has been lost. In budgeting, however, some time must be allowed for locating respondents and for pauses between interviews to give the interviewer a chance to catch her breath and organize notes.

A major cost of phone interviewing are the telephone toll charges. One reason for the great increase in phone interviewing over the past decade has been the availability of WATS lines, which permit unlimited calls after monthly payment of a fixed rental charge. Since the rental charge is large, it would not pay to install a WATS line for a small one-time survey. Where a WATS line is already available, however, or can be justified because of the multiple uses that will be made of it, the individual costs per phone interview will be small. These costs should be included, however, in budgeting phone procedures.

Face-to-face methods are the most expensive because the costs include not only direct interviewing time but all the travel costs involved in locating a respondent. Using optimum procedures, these travel costs typically account for more than half of all interviewing costs. Usually survey organizations pay interviewers for face-to-face interviews from the time they leave their homes until they return home, as well as a mileage allowance for using their cars. Not only is face-to-face interviewing the most costly of the procedures, it is also the most difficult to budget accurately.

When one looks at the quality dimension, however, the more costly procedures are also the most versatile and, in most cases, probably give the highest-quality data. There are three major sources of error in a survey: (*1*) *sampling variability, generally called sampling error, which depends on the sample size and design;* (*2*) *sample biases which are a function of how well the study design is executed; and (3) response effects which are the differences between reported and "true" measures of behavior, characteristics, or attitudes.*

Given a fixed amount of money, the cheapest procedure per case will, of course, allow one to survey a larger sample. The sampling error for the same expenditure of funds is lowest for a mail survey, since the sample size is largest. The sampling error is largest for face-to-face interviews.

Sampling error, however, is generally the smallest of the three components of error. Sample biases, which may often be several times larger than sampling errors, are smallest for face-to-face interviewing and largest for mail questionnaires. There are two major kinds of biases in mail surveys. The more general one is related to education. On any kind of survey, highly educated respondents are more likely to cooperate than are poorly educated ones, for whom filling out forms is a more difficult and frightening task. The more specific biases on mail questionnaires are due to the respondents' interest in the study's subject matter or loyalty to the study's sponsor. A mail survey of the readership of a magazine will get a higher response from those who find the magazine most useful. A survey on any controversial issue will get more returns from those who are strongly for or against a given position, and less response from those in the middle. Mail surveys, because they take so long, are also not appropriate when a quick answer is needed because attitudes are changing rapidly, as during a domestic or foreign crisis.

Phone and face-to-face methods avoid the biases of mail procedures but are subject to biases of their own. Phone interviews are limited to those respondents with telephones and thus are biased against lower-income households without phones. In some suburban areas, these biases are trivial, with better than 95% of households having telephones. Even in poor areas of most large U.S. cities, better than 80% of households have telephones; but, in poor rural areas, the fraction of households with phones may be as low as 50–60% and the biases become of major importance. In countries other than the United States, where phone availability is far lower, phone interviewing is used only for special samples.

The major bias connected with face-to-face interviewing is the nonavailability of the respondent when the interviewer comes to the door. This can be a troublesome problem particularly in large cities, but it can be solved with sufficient cost and effort.

There are no generalizations that can be made about the relation of response effects to the method of data collection. For some kinds of threatening information, the less personal mail and phone methods appear to be superior to face-to-face interviewing. For questions that require probing and large quantities of detailed information, face-to-face methods are better. Considering all forms of error combined, the most expensive face-to-face procedures usually have the smallest total error.

1.9 THE USES OF COMBINED METHODS

Given the cost advantages of the mail procedures and the quality advantages of the personal methods, how can one decide the method to use? The decision will depend on the available funds and the quality of the sample required. A very useful procedure that may not occur to the novice is the use of combined methods. These combined methods use the optimum sampling procedures discussed in Chapter 6 and are more complex than using only a single method. The combined sample, however, may be far more efficient—that is, for the same funds, it yields a much larger and higher-quality sample.

Typically, one starts with the cheapest procedure, the self-administered mail questionnaire if the population is appropriate, or a telephone interview. A subsample of respondents who cannot be reached by cheaper methods are then interviewed by the more expensive procedures. The proportion subsampled depends mainly on the ratio of costs of the different procedures, as in Example 1.11.

Example 1.11 Doctors and Cigarette Smoking

A study was conducted for the U.S. Public Health Service to determine the current smoking habits of physicians after information had been released on the relation between smoking and lung cancer. Face-to-face interviews were considered impractical because of doctors' schedules. Instead, the intitial contact was made with a mail questionnaire followed by two additional mailings to doctors who did not cooperate. This resulted in about slightly fewer than half of the doctors returning a questionnaire. (Because of their tight schedules and the heavy amount of mail they receive, doctors do not respond as well to mail questionnaires as do other professionals.)

A subsample of 40% of the noncooperators was selected for long-distance telephone interviewing. Initial calls were made to find a convenient time to call for the interview. From this process, about 80% of the doctors in the subsample were interviewed. The final reported results were based on a weighted sample of the two groups, with the doctors in the telephone sample receiving a weight of 2.5 to account for the fact that only 40% had been selected. Additional details on this sample design are given in Example 6.10.

1.10 SHOULD AN EXISTING SURVEY ORGANIZATION BE USED?

The decision by a researcher to use an existing organization or to do the field work himself should depend on the scope of the project, and the type of population. For very small projects with limited funds available, do-it-yourself may be the only alternative. For larger studies, it may at first seem cheaper for the researcher to hire, train, and supervise a staff of interviewers, but this will seldom be the case if one considers the time and travel costs required. For large national population studies, the initial costs of hiring and training an interviewing staff may exceed $100,000. These costs would overwhelm most project budgets unless they are shared by a large number of different projects; then the costs to any one study become reasonable.

If one wishes to use an existing organization with a national field force of interviewers, one would not select a sample independently, since it would be unlikely that any organization had trained interviewers in the newly selected locations. Rather, one would first select the organization and use the national sample of that organization, or some modification of it.

1.11 UNUSUAL POPULATIONS

If the population to be studied is a general one, such as all adults or households in the United States, there is no question that it is more efficient to use an existing organization. For special samples, the decision is more complex. For those that can be surveyed by mail or existing WATS lines, the advantages of a national field force are irrelevant. If the mailing is small, it can be handled using manual procedures. If the mailing is large, it still may pay to use an outside organizing with the machinery and experience in mailing and in keeping track of returns.

Suppose one wishes to do face-to-face interviewing with special groups, such as Spanish-American War veterans, Catholics, or judges. These groups may be located in areas where national field organizations do not have interviewers. Even here, however, the use of an existing organization in combination with the hiring of new interviewers is more efficient than training a whole new staff.

Example 1.12 Spanish-American War Veterans

A study of 1500 surviving Spanish-American War veterans, whose average age was 85 at the time of the study, was conducted to determine medical care utilization of this group. These veterans are distributed geographically in approximately the same way as the total population, with one major exception—there is a very heavy concentration in Florida. When the study was conducted, a national field force of interviewers was used, while in Florida new interviewers were hired. The procedure is basically a stratified sample with Florida as a separate stratum (see Chapter 6). (It should be noted that both interviewers and respondents enjoyed the interview. Interviewers reported that their very old respondents were lonely and enjoyed the chance to talk about themselves with sympathetic listeners.)

Many special groups are distributed geographically in about the same way as the general population, such as judges or participants in various government programs. Some of these programs, although national in scope, place particular emphasis on certain groups, such as Appalachian whites, Indians, or Mexican immigrants. Each of these special groups would form a stratum that required special sampling, as in Example 1.12, to supplement a national sample.

In some special cases, the geographic location of the population precludes the use of a national field force and requires that a new sample be selected and all new interviewers be hired. This would be the case if one were trying to study the assimilation of Cuban refugees into the U.S. society, or to compare the behavior of American Indians who live on and off the reservations. In these cases, not only would ethnic interviewers be necessary, but also the population is located mainly in places that are not usually part of a national sample.

1.12 SAMPLING FROM LIMITED GEOGRAPHIC AREAS

The growing ability of both public and private state and local survey research organizations is an indication of a need for local surveys. Many public policy decisions can be made only at the local level, such as property taxes, the location of new highways, and the adequacy of existing public schools, police and fire protection, and garbage removal. At the state level, it may be vital to know public attitudes toward increased taxation as well as toward how state resources should be allocated.

It is important to recognize that there are no basic distinctions between national and state or local samples. To do a careful study of Texas or California or the Chicago and New York metropolitan areas requires the same procedures as selecting a national sample of the United States. The cost of a local sample may be less, primarily because travel costs are lower in any geographically limited area.

Many local and state samples will be smaller than national studies because the national policy decisions involve far greater risks in the allocation of government spending, but this is not always the case. Important state problems for larger states like New York or California may require larger sample studies than less sensitive national studies that are more research oriented.

The availability of local field organizations and the relatively lower cost of local field work may tempt the researcher to do a local study when the population of interest is really national. This decision is rationalized on the grounds that the state or local area is a microcosm of the entire country. Thus, one hears the argument that Illinois or Ohio or New York State have both urban areas and rural areas, rich communities and poor ones, and that within their borders they reflect national differences.

It is not difficult to poke holes into this sort of argument. There are major differences in income, occupation, and attitudes between different parts of the country that cannot be ignored if one is attempting a national study. On the other hand, a state or local survey may be appropriate during the early stages of a study when one is testing methods and developing hypotheses.

Some experiments are so costly and complex that they can be done only in one or a few local areas, although the researchers recognize that this will limit the generalizability of the results. Examples of this are studies like the income maintenance experiment conducted by Watts (99, 100) and the housing assistance experiment conducted by Lowry and the Rand Corporation (47). In both these experiments, substantial amounts of money are provided to poverty families to measure the impact of these grants on employment behavior and housing choices. The samples must be kept small and local to keep the experimental costs within reason, even though both these projects have multimillion-dollar annual budgets. In addition to financial considerations, the political implications of these experiments require that cooperation be obtained from appropriate officials before the experiment is begun. This is a long and sensitive task that would be impossible to accomplish in many places simultaneously.

1.13 SELECTING PLACES FOR EXPERIMENTS

The differences between states and cities in their legal systems, welfare programs, public services, income levels, and other variables sometimes may be used to strengthen the experiment if only a limited number of places can be afforded. Instead of choosing convenient places, the researcher decides on the places on the basis of those variables most closely related to the purposes of the experiment.

Example 1.13 Housing Assistance Supply Experiment (47)

A decision was made by the Department of Housing and Urban Development to fund the Rand study, described in the previous section, in two metropolitan areas with populations under 250,000. The purpose of the study was to measure the effects of housing grants on the supply of housing. Aside from size, it was determined that the two variables most critically related to housing supply were the growth rate and the percentage of blacks in the central city. A four-cell classification was used: slow central-city growth rate (6.9% or less from 1960 to 1970), low percentage of blacks (10.8% or less); slow central-city growth rate, high percentage of blacks (more than 10.8%); fast central-city growth rate (7.0% or more), low percentage of blacks; fast central-city growth rate, high percentage of blacks. The greatest share of the metropolitan population falls in the slow-growth–high-black category, so one site was selected from this group. The other site was selected from the fast-growth–low-black category, for the greatest contrast.

Systematic screening procedures eliminated places that did not really fit their assigned categories (growth rate distorted by annexation), that would be politically difficult (SMSAs that straddle state lines), or that were unsuitable because of some unusual characteristic (a large military or college population). The final choices of sites depended on obtaining cooperation from the political leaders in the area. The entire process is far from being a representative sample of all metropolitan areas, and there is no way to compute estimates of sampling variability for the broader population. Nevertheless, the data generated by this study should be more useful and convincing to political decision makers than a sample of sites selected only for political reasons or convenience.

1.14 THE USE OF LOCAL AREAS FOR NATURAL EXPERIMENTS

It has often been pointed out by political scientists that a federal form of government provides the opportunity for natural experimentation. If one wishes to measure the effect of any legislation or public program, one can compare the results in states or cities where the legislation is in effect to the results where it is not. Thus, the effects of differing welfare payments or medical programs can be compared between states to see if infant mortality or unemployment or any other dependent variables differ.

The difficulty, of course, is controlling for other variables. Those states with the lowest welfare payments are mainly states with lower per capita income in the South. It may not be possible to separate the effects of the public policies from the other effects. If it is possible to find places that are similar on some major demographic variables but different on the variable being studied, a useful natural experiment is possible. While the results will always be inconclusive, since not all differences can be controlled, the natural experiment is sometimes the best alternative to doing nothing or designing a controlled experiment that is politically impossible.

1.15 SUMMARY

In this chapter, a number of different samples have been discussed, ranging from very high to poor quality; each of them can be appropriate under certain conditions. A small local study is appropriate for a researcher with limited resources who is just beginning to explore an area. The same design would not be appropriate in the later stages of research or for major policy decisions nationally. On the other hand, no sample is without some limitations. To insist on perfection or on very high quality in all situations is to misunderstand how knowledge grows. Where no searchlight is available, it is better to light a candle than to curse the darkness.

Even if resources are limited, it is important to define the population carefully and to consider the design of an ideal sample. If this is done, one will have a realistic notion of the quality of the sample and its limitations. The alternative of defining the population to fit the sample should be shunned.

1.16 ADDITIONAL READING

I strongly suggest that some of the examples given in this chapter be read in their original versions, with particular attention to the technical appendices that discuss the sample design. (The complete reference data will be found at the back starting with page 235.) In addition, there are many other studies that will include discussions of the sampling designs used. You may wish to consult the following bibliographies:

Bureau of the Census Catalog (89)
NORC *Bibliography of Publications* (53)
University of Michigan, Institute for Social Research *List of Publications* (37)
Bureau of Applied Social Research *Bibliography* (10)
University of California, Berkeley, Survey Research Center *Publications List* (83)

For the reader who wants a general discussion of survey methods, the second edition of *Survey Methods in Social Investigation* by Moser and Kalton (52) is highly recommended. Some interesting and useful examples of sampling for business problems are given by Deming (18).

In addition to the technical journals in their own fields, readers will find useful examples of sample designs in the following:

Public Opinion Quarterly
Journal of the American Statistical Association
Journal of the Royal Statistical Society, Series A

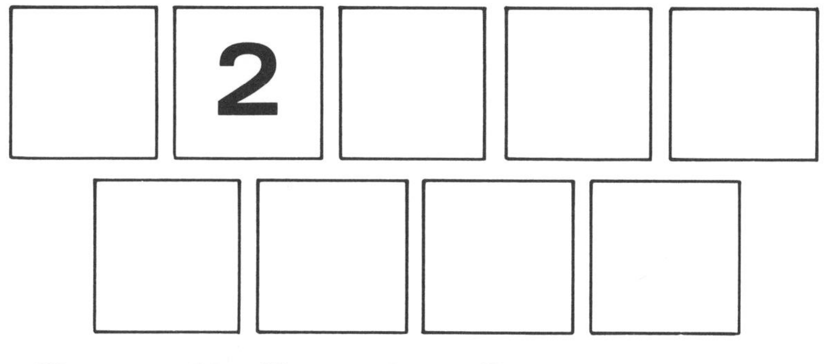

Small-Scale Sampling with Limited Resources

2.1 INTRODUCTION

Most social scientists who have no training in sampling get their ideas on what to do from reading other people's research in scientific journals. The reviewers of journal articles, however, are themselves not always experts in sampling and are primarily concerned with the substance of the article and the analytic procedures used. Thus, the quality of sampling in published studies varies enormously. Many of the studies reported are based on small-scale samples that have serious limitations, which are sometimes recognized and often ignored.

In this chapter, we look at studies selected from recent issues of the *American Sociological Review,* the *American Journal of Sociology,* and *Public Opinion Quarterly,* and comment on the quality of sampling used in them. This is by no means a random sample of studies. Very large national and cross-national studies have been omitted since they are discussed later in this book. Where multiple studies were found using the same sampling methods, only one or two were chosen for illustration. Also, while some of these studies have inappropriate sample designs, the editorial screening process has prevented most of the studies with really bad samples from ever seeing the light of day. The purpose of this chapter is to alert researchers with limited resources to procedures for improving the quality of their samples as well as to suggest criteria for reader evaluation of sample credibility of published research.

In criticizing some of these studies, we are not concerned with their theoretical or analytical procedures. A poor sample design should not lead the reader to believe that the findings of a study are necessarily invalid or that contrary results are indeed correct. Rather, concern about sampling methods should lead to reduced credibility of the findings and an increase in the uncertainty about their generalizability. The refutation of study findings must come from other studies with greater credibility.

2.2 A CREDIBILITY SCALE

To formalize the notion of credibility and organize the discussion in this chapter, a credibility scale has been developed for judging small-scale samples. The credibility scale includes the factors that samplers would generally consider in looking at a sample: the generalizability of the findings, sample size, the execution of the sample design, and the use of the available resources. The items included in the scale and the weights assigned are given in Table 2.1.

The weights assigned and the scoring of individual studies are personal judgments, so that different samplers might assign different weights and might rank samples somewhat differently from the way they are listed here. These weights should not be used uncritically to distinguish between samples with similar levels of quality. Nevertheless, readers should be able to detect the differences between the best and worst sample designs and be able to apply them in their critical reading as well as in their planning and write-ups of their own sample designs. It should also be noted that even the best of the designs discussed in this chapter have serious flaws and do not compare to the larger standard samples discussed in later chapters.

GEOGRAPHY

Before turning to a discussion of the specific studies, a brief discussion of the factors in the scale may be helpful. Greatest emphasis is placed on how well the data may be generalized. Unless one is dealing with a small special population in a single location, a limited sample does not usually represent the total universe. If one observes the same results in several locations with widely differing populations, however, one has a great deal more confidence in their generality than if the sample is only of a single location. The greatest relative increase in quality is achieved by increasing the number of locations from one to two, and comparing the results from the different sites. Combining the results of several locations could conceal important site effects and should be done only after a careful analysis has indicated no significant site differences.

The researcher with limited funds may feel that control of the field work and quality of the data collection will be improved by limiting the sample to a single

Table 2.1
Credibility Scale for Small Samples

	Score
A. Generalizability	
1. *Geographic spread*	
Single location	0
Several locations combined	2
Several locations compared	
Limited geography	4
Widespread geography	6
Total universe	10
2. *Discussion of limitations*	
No discussion	0
Brief discussion	3
Detailed discussion	5
3. *Use of special populations*	
Obvious biases in sample that could affect results	−5
Used for convenience, no obvious bias	0
Necessary to test theory	5
General population	5
B. Sample size	
Too small, even in total, for meaningful analysis	0
Adequate for some but not all major analyses	3
Adequate for purpose of study	5
C. Sample execution	
Poor response rate, haphazard sample	0
Some evidence of careless field work	3
Reasonable response rate, controlled field operations	5
D. Use of resources	
Poor use	0
Fair use	3
Optimum use	5
Total points possible	35

location, but this assumption should be examined very carefully. Frequently, it will be found that tighter control can be maintained over small crews in several locations than over a larger interviewing group in one location, although it may require more effort by the field supervisor or project director.

Another alternative, which is observed in Example 2.1, is for two researchers in widely scattered locations to collaborate. The results obtained by combining resources are substantially better than the sum of two separate studies. Still another helpful method for increasing sample credibility is to compare the results of a study to those of earlier studies. If the results replicate those of earlier studies, both the old and new studies gain in credibility, even if the methodologies and questionnaires differ. If, however, the results of a study contradict the results of earlier ones, the researcher is faced with serious problems of deciding whether the differences are caused by sample differences, different measurement procedures, or something else.

DISCUSSION OF LIMITATIONS

A careful discussion of the study's sample limitations is useful, especially for readers with limited sampling backgrounds. Thus, a study that carefully states and explores its possible sample biases gains rather than loses credibility.

As an example consider the following excerpts from Lenski's *The Religious Factor:*[1]

> The study was carried out in the Midwestern metropolis of Detroit, fifth largest community in America today, and probably eleventh largest in the world. Here, by means of personal interviews with a carefully selected cross-section of the population of the *total* community (i.e., suburbs as well as central city), we sought to discover the impact of religion on secular institutions. Strictly speaking, the findings set forth in this volume apply only to Detroit. However, in view of the steady decline of localism and regionalism in America during the last century, it seems likely that most of these findings could be duplicated by similar studies in other communities. This is a matter to which we shall return later in this chapter. . . .
>
> In its economics, politics, ethnicity, and religion, Detroit most closely resembles Cleveland, Pittsburgh, Buffalo, and Chicago. In common with these communities, Detroit is noted for heavy industry, high wages, a large industrial population, a large proportion of eastern European immigrants of peasant background, and a rapidly growing Negro minority recently arrived from the rural South. Among the major metropolitan centers it bears least resemblance to New York and Washington, both of which differ markedly in terms of economics, ethnicity, and religion, and, in the case of voteless Washington, in terms of politics as well.
>
> Despite these local peculiarities it seems probable that most of our findings in Detroit can be generalized and applied to other major metropolitan centers throughout the country, with the possible exception of the South. This appears likely for two reasons. In the first place, the issues we investigated are basically national in character, and not local. Advances in transportation and mass communication mean that people all over the country are nowadays subject to similar pressures and influences. Local and even regional peculiarities have been progressively eroded. More and more the nation is becoming a political, economic, religious, and social unit. Secondly, Americans are becoming more and more mobile. *Of those now living in greater Detroit, nearly two thirds were born elsewhere.* More than half were born outside of Michigan, and therefore outside the sphere of Detroit's direct influence, and within the orbit of some other metropolitan center. This constant movement of population

[1] Gerhard Lenski, *The religious factor* (Garden City, New York: Doubleday, 1961), pp. 1, 33–34. © 1963 by Doubleday & Company, Inc.

> also hampers the development of regional peculiarities, and promotes the homogeneity of the national population.
>
> In the last analysis, however, the only sure test of the generalizability of the findings of a study based on a single community can come from similar studies conducted elsewhere. For this reason, throughout this book references will be made (usually in footnotes) to earlier studies which have dealt with similar problems elsewhere. In this way the reader will be better able to judge to what degree the findings of this study are unique to Detroit, and to what degree they may apply to other communities.

While the extended discussion of sample limitations possible in a book or monograph must be condensed in a journal article (if not by the author, then by the editors) some discussion of the critical differences between the sample and universe should be included.

USE OF SPECIAL POPULATIONS

The use of special populations may sometimes be a powerful tool for testing a theory. In a study of the socialization of children, samples of school children are highly appropriate. For testing organizational effects on managers or workers, the firm is the logical place to begin. In Example 2.6, cadets at a military academy are used to study professional socialization.

In some cases the use of a special population may lead to obvious or potential biases. Thus, the use of college students to represent the total population leads to major education and social class biases. In addition, the authority relation between the students and the researcher may be such that response effects are greatly magnified.

A common use of special populations is in secondary analyses when data initially collected for one purpose are reanalyzed for a different purpose. If the initial data were from a general population sample, then, of course, there are no problems. Potential biases arise when a special population is treated as a regular population sample in the reanalysis. This procedure is sometimes justified because it makes very efficient use of limited resources; in this case, it is especially important that the sample be critically examined by the researcher for all possible biases.

SAMPLE SIZE

A full discussion of sample size determination is given in Chapter 5. Here we only mention that the adequacy of the sample depends on the details of the analysis. Few studies seen in the literature have samples that are too small when only the total sample is used. For most analyses, however, breakdowns of the sample are required; for many breakdowns, the observed samples are inadequate.

A general rule is that the sample should be large enough so that there are 100 or more units in each category of the major breakdowns and a minimum of 20 to 50 in the minor breakdowns.

SAMPLE EXECUTION

The quality of a sample depends not only on its design but also on its execution. Low cooperation rates may indicate sloppy field work and lack of follow-up procedures. A frequent example of this is seen in mail surveys that use a single mailing and obtain low cooperation rates when additional mailings could increase cooperation to the generally accepted level of about 80%. The biases in mail samples are toward those respondents with more education and those who are most interested in the topic.

Even worse are personal samples in which the interviewer is allowed to select the respondents or households to be interviewed. Here no measure of cooperation is possible and the biases are likely to be toward the most accessible respondents. These are more likely to be women, unemployed, middle-aged or older, and middle-class.

USE OF RESOURCES

Although this factor is independent of judgments about the absolute quality of a sample, it seems appropriate to consider how well the sample was designed and executed with the resources that were available. Several of the examples to be given later report studies that were conducted in response to a specific news event. In these cases, the researchers rushed into the field with very limited resources. If they had waited to obtain funds and select a careful sample, the timeliness of the research would have vanished. Even here, however, some sampling methods are far better than others. A quick phone sample, for example, is far superior to street-corner interviewing, since it is far less biased and no more costly.

The use of natural clusters, such as classrooms when the study deals with children or college students, is also an efficient use of limited resources. On the other hand, if the study is to be conducted by mail, heavy clustering is an inefficient use of resources, since it reduces the generalizability of the results without reducing costs.

2.3 EXAMPLES

The examples listed here are in decreasing order of credibility, based on the credibility scale of Table 2.1 and on my judgment. A brief discussion of the aim of the study and the sampling method used is given, as well as the scores on the individual components of the scale. For additional information about the studies, readers are urged to consult the original articles, which should be readily accessible.

Example 2.1 "Effects of Vertical Mobility and Status Inconsistency: A Body of Negative Evidence" (38) and "Community Rank Stratification: A Factor Analysis" (6)

Both these studies are based on the same sample of six communities, three in Indiana and three in Arizona. Male heads of households were drawn randomly from the street address sections of city and suburban directories. The sample size was 686 males in Phoenix and between 300 and 400 in the other cities. In Indianapolis, the interviewing was part of the Indianapolis Area Project, a training program similar to the Detroit Area Study. In the other cities, interviewing was done by Elmo Roper Associates, a well-known research firm.

The first study attempted to determine whether dimensions of social rank combine additively, or interactions appear to support the notions of status inconsistency or vertical mobility. The results favored the additive models. The second study, a factor analysis of rank measures, suggested that stratification systems vary by community context.

Credibility Score. 31/35 = .89.

Generalizability 6. The use of six locations widely separated and of different sizes is very useful. As Jackson and Curtis (38) put it:

> Analyzing our problem in several rather different communities allows us to estimate whether mobility and/or inconsistency effects are more or less general, or whether they appear only in certain social settings. It also allows us to see which effects do not replicate across cities in any fashion and hence should possibly be labeled chance fluctuations [p. 702].

The results reported in the first study (38) are used to disconfirm a theory. As is well-recognized, the requirements for confirming a theory are substantially stronger than those for disproving one. Here the absence of positive evidence of status inconsistency and vertical mobility in six different communities would lead most readers to accept the research and reject the theory.

Discussion of Limitations 5. A careful discussion of the selected communities.

Use of Special Populations 5. The universe is limited to male household heads, which seems appropriate given the aims of the study. Households are selected at random from city directories (see Chapter 3).

Sample Size 5. It is clear that the sample sizes here are ample for the analysis.

Sample Execution 5. Although the completion rates are not given, and it would have been useful to have them, all evidence is that the study was done very carefully.

Use of Resources 5. This sample has two excellent examples of the careful use of resources. First, it combines research with training of students; second, it combines the resources of researchers in Indiana and Arizona.

While this is not a national study, some readers may feel that the sample size is too large for this study to be considered small-scale. The quality would not suffer very much, however, if the samples were considerably smaller, or if only four instead of six communities had been used. Thus, many of the techniques seen here could be used by researchers with more limited funds.

Example 2.2 "Social Position and Self-Evaluation: The Relative Importance of Race" (101)

This study evaluates the effects of race, sex, city, age, education, marital status, and employment on self-esteem and stress: Race has minimal effects when other variables are controlled. The study was conducted with 362 blacks and 350 whites in Nashville, Tennessee, and 215 blacks and 252 whites in Philadelphia, Pennsylvania.

Employing 1960 census information and any information available on subsequent neighborhood change, residential areas in both cities were selected which were thought to hold lower-, working-, and middle-class blacks and whites. Within each residential area, blocks were randomly chosen. Specific dwellings were selected by systematically interviewing in every fifth dwelling unit. Within each dwelling unit, the interviewer attempted to interview the head of the household, but interviewed some second adult when it became clear that the household head was unavailable. The nonresponse rate, given three call-backs, was under 5% in each city. An effort was made to match race of interviewer with race of respondent, but approximately 35% of black respondents were interviewed by white interviewers. An analysis indicated no effects of the interviewer's race on the results related to stress.

Credibility Score. 30/35 = .86.

Generalizability 5. Here is another example of researchers, one in Philadelphia, the others in Nashville, combining their resources to give the results in two locations. While slightly less convincing than the results in six locations, these results are far better than if only one city had been used. Note also that this research was intended to disconfirm a theory, so fewer locations were needed.

Discussion of Limitations 5. There is a discussion of the differences between the two sites and the general effects of city and region.

Use of Special Populations 5. It is a general population sample with the head of the household interviewed.

Sample Size 5. The size is clearly adequate.

Sample Execution 5. The sample was carefully done with a very low noncooperation rate.

Use of Resources 5. The resources for this study were obviously less than for Example 2.1, but they were well utilized by combining researchers from two locations.

Example 2.3 "The Development of Political Orientation among Black and White Children" (61)

Black and white school children in Grades 4–12 were studied in four urban areas in Illinois. Self-administered questionnaires were obtained from 2365 students, split equally between black and white and male and female. Of the sample, 50% was from an inner-city, lower-class, black, public school system; 10% from an inner-city, middle- to upper-class, integrated school. The last two areas each contributed 20% of the sample, with one constituting a lower-middle to working-class public school system and the other a middle-class public school system, both being principally composed of white students.

Credibility Score. 29/35 = .83.

Generalizability 5. The sample benefits from four quite different areas, but the generalizability is limited because all students are from Illinois. Particularly in studying attitudes of black children, comparisons between the North and South could be important.

Discussion of Limitations 5. There is a careful comparison of the results of this study with the results of other studies, indicating some similarities but also some disparities.

Use of Special Populations 5. The use of school children here is necessary for the analysis.

Sample Size 4. Although, in total, this sample is very large, most of the analyses are carried out controlling for race, occupation of chief wage-earner (blue collar, white collar), and grade in school (4–6, 7–8, 9–10, 11–12). Some of the analyses of differences are based on rather small samples.

Sample Execution 5. Except for children absent on the day the form was administered, there was no noncooperation once the school agreed to cooperate.

Use of Resources 5. There was a very efficient use of limited resources, using self-administered forms in a classroom and using graduate students in the Survey Research Practicum at the University of Illinois, Urbana, to collect the data.

Example 2.4 "Employment Opportunities for Blacks in the Black Ghetto: The Role of White Owned Businesses" (1)

A study of black- and white-owned businesses in black ghettos in Boston, Chicago, and Washington, D.C., indicates that white-owned businesses are much larger than black businesses, dominating the labor market of the ghetto, and that white owners are much more likely to hire "outsiders" and whites. For comparison, ghetto areas are compared to nonghetto areas in the three cities. Of the total of 512 business sites, interviews were obtained from 431, giving a completion rate of 84%. This study was a follow-up to an earlier study of crime and law enforcement in the same neighborhoods.

Credibility Score. 28/35 = .80.

Generalizability 6. While three sites were used, the reader is troubled by the lack of locations in the West and South, although it may be argued that Washington, in some ways, is a Southern city.

Discussion of Limitations 5. There is a detailed discussion of the six neighborhoods. The fact that patterns of ownership and operation are quite comparable in all the neighborhoods studied is used to defend the generality of the findings, although occasional differences are observed.

Use of Special Populations 5. This is a study of small businesses, which is the population used.

Sample Size 3. The sample sizes are adequate when all cities are combined, but they become small in the individual neighborhoods, especially the black-owned businesses. There are only 18 black-owned businesses in the two neighborhoods in Boston, and 25 in Chicago. Since all the businesses in the neighborhood were sampled, the only way to increase the sample size would be to add more neighborhoods.

Sample Execution 5. There was a high cooperation rate within neighborhoods.

Use of Resources 4. Although the use of neighborhoods used earlier for another purpose is efficient since the location of businesses already had been done and interviewers were available, the firms chosen in these neighborhoods are not evenly distributed between black and white ownership. This results in a sample of 377 white-owned and 134 black-owned firms. For comparison, it would have been better to increase the number of black-owned firms and reduce the number of white-owned firms until the sample sizes in the two groups were about equal. This would have required additional neighborhoods plus subsampling of white-owned firms.

Example 2.5 "The Structure of Scientific Fields and the Functioning of University Graduate Departments" (45)

Four academic fields—physics, chemistry, sociology, and political science—were studied and it was found that the relatively high paradigm development in the physical sciences facilitated agreement over field content as well as greater willingness to interact with graduate students than in the social sciences. A stratified random sample of 20 university departments in each field was chosen. The sample did not represent all graduate schools, only those listed in the Cartter Report of high-quality schools; within departments, all faculty members were surveyed. The overall response rate was 51% and a total of 1161 responses were received.

Credibility Score. 26/35 = .74.

Generalizability 6. There are two major concerns with the generalizability of this sample, although it involves 80 randomly selected departments. The first is the omission of schools not named in the Cartter Report. There is no discussion of why this procedure was adopted rather than using all schools that offer graduate training. The second is the selection of the fields. While there is a careful discussion of why these fields were selected, one might wonder if the results would have been the same had the biological sciences or other physical sciences, such as astronomy and geology, been included.

Discussion of Limitations 3. There is a brief discussion of possible sample biases indicating no differences by age or rank of respondents.

Use of Special Populations 5. The use of faculty members is appropriate for this study.

Sample Size 5. The sample is large enough for comparison between the four fields.

Sample Execution 3. The low response rate in this study is due to the fact that only a single mailing was used. Two follow-up mailings probably would have increased the cooperation to 80% or higher and removed concerns that respondents most interested in the topic were more likely to respond.

Use of Resources 4. The use of mail for contacting faculty members is an efficient way to gather data when there are limited resources. It would have been better to put some of the funds into follow-up mailings to reduce bias, even if this meant that fewer schools could be sampled.

Example 2.6 "Power and Ideological Conformity: A Case Study" (29)

This is a study of professional socialization at Britain's Sandhurst Military Academy, indicating that conformity to the staff's goals is achieved. The sample size of 883 comprised 92% of the cadets at the academy during 1967.

Credibility Score. 26/35 = .74.

Generalizability 1. Although this is only at a single location, some effort is made to generalize by discussing other studies. If resources were available, one could attempt to extend this study to other military academies in the United States or to officer candidate schools and other military training programs.

Discussion of Limitations 5. Given the limited nature of this study, the discussion of limitations is excellent. In addition to calling it a case study, the author concludes the paper with the following remark:

> It must be noted that these findings stem from a case study. While the findings presented here seem reasonable, they are nevertheless based on limited evidence. If the study of socialization, and particularly socialization taking place within organizations, is going to proceed further, the time may have come to manipulate the variables isolated here in the laboratory. If the laboratory is too artificial a setting, then future researchers should make sure that the variables described in the literature are systematically manipulated. Only then can reasonable findings become scientific [p. 362].

Use of Special Populations 5. Military academies are appropriate settings for studies of power.

Sample Size 5. Since all cadets were included, the sample size is clearly sufficient for this case study. For broader generalization, the sample could be increased by using more sites.

Sample Execution 5. The 92% cooperation rate is very good and indicates careful field work and follow-up procedures.

Use of Resources 5. The use of self-administered forms for all cadets indicates a careful use of resources.

Example 2.7 "Interracial Public Housing in a Border City: Another Look at the Contact Hypothesis" (27)

Black and white housewives were interviewed in racially segregated and desegregated projects in Lexington, Kentucky, to determine the extent to which engaging in equal-status interracial contacts was related to racially tolerant attitudes. A total of 168 housewives were selected from six projects, with the cooperation rates ranging from 75% to 89%. The findings supported the contact hypothesis for lower-income white housewives, in contrast to some earlier studies, but failed to support the contact hypothesis for black women.

Credibility Score. 25/35 = .71.

Generalizability 4. Although the study was conducted in only a single city, it was conducted at six different sites which are compared. In addition, there was a careful effort to use questions and scales that had been used in earlier studies, and to compare the results.

Discussion of Limitations 5. The detailed discussion of the sample and its limitations, as well as the comparison to earlier studies, is one of the strongest features of this study.

Use of Special Populations 5. The use of residents of public housing is justified by the theory to be tested, although the analysis at times veers away from current housing to previous interracial contacts.

Sample Size 3. The sample is barely adequate for comparisons between black and white housewives and becomes too small when only those who live in interracial projects are analyzed.

Sample Execution 5. The cooperation rates, along with the detailed description, indicate that the field work was carefully done. Black housewives were interviewed by two black interviewers and white housewives were interviewed by the author.

Use of Resources 3. Given the limited resources, the small sample is understandable. What is unclear is why one-third of the sample came from segregated projects. These households are excluded from the crucial analyses, leaving only about 50 white and 50 black households. The analysis might have been broadened to include these households as having zero interracial contact; or, if the published table had been conceptualized before the study, segregated projects could have been omitted and the resources used for increasing the remaining sample.

Example 2.8 "Managerial Mobility Motivations and Central Life Interests" (32)

Using 489 middle managers and specialists in seven American industries, the study relates "career anchorage points" and central life interests. The data indicate that, regardless of age, education, level of labor force entry, and present position, "upwardly anchored" managers and specialists are more work-oriented.

Credibility Score. 24/35 = .69.

Generalizability 3. The study was conducted in seven firms: two steel companies; a vertically integrated lumber and paper firm; a maker of paints, waxes, and wallpaper; a firm producing large plastic signs; a big city bank; and a mattress and box spring manufacturer. Although this is quite an assortment, the criteria for choosing these specific firms are not given; we can only speculate that possibly the investigator knew someone in each firm. Also, the data for firms are never shown separately so that one can determine something about the variability among firms.

Discussion of Limitations 4. Although a discussion of how the firms were selected is omitted, there is a discussion of the selection of individuals within firms and a careful discussion of other studies on the same topic.

Use of Special Populations 5. The use of middle-management executives and specialists is necessary to test the theory.

Sample Size 3. The total sample size is adequate, but the interesting comparisons are between "upward career anchorages" and the rest. There are only 66 respondents in the "upward" category, making many of the analyses very shaky.

Sample Execution 5. A mail-back system with two follow-ups produced an overall cooperation rate of 85%, with the lowest rate from any company being 82%. This is very satisfactory.

Use of Resources 4. The use of mail is an efficient procedure for contacting business executives once they have been located. The major problem with this study is that only 13% of the sample fell into the most interesting category, "upward career anchorage." A possible procedure for increasing the sample size of this group would have been to use some sort of screening questionnaire with a larger initial population and then to take all those with "upward career anchorages" and a subsample of the remainder. While this general procedure is usually very efficient, it might have been impossible to do any screening for this study.

Example 2.9 "Race, Class, and Consciousness" (36)

This study relates race, class, size of birthplace, and age to class consciousness, race consciousness, and interest in politics. The data come from a sample of 1870 whites and 434 blacks in four cities in Connecticut. The findings indicate that both race and class are important and reinforce each other in their effects on feelings of exploitation or privilege, but that race is somewhat more important.

Credibility Score. 24/35 = .69.

Generalizability 2. The major problem with generalizing the data from this study is that all four cities are combined so that there is no way to tell if there are variations between cities. Also, blacks in Northern industrial cities may not feel the same as blacks in the South and in rural areas.

Discussion of Limitations 5. There is a careful discussion of the sampling and field methodology as well as a detailed comparison to earlier research.

Use of Special Populations 5. This is probability sample of white and black households in the selected cities.

Sample Size 4. While the sample is large enough for classification by a single variable within race, it is too small among blacks for any further breakdowns.

Sample Execution 5. The completion rate varied from 85% to 89% indicating, as does the discussion, that the field work was done carefully.

Use of Resources 3. For the purposes of this study, the sample would have been more efficient if half the sample were white and half black. This study, however, is one of a series based on results from the same locations; it may be that the other studies did not require as large a sample of black households.

Example 2.10 "How Fast Does News Travel?" (73)

The shooting of Governor George Wallace on May 15, 1972 provided an opportunity to measure how fast nationally significant news travels. Interviews were conducted by phone in New York City between 5:00 and 10:00 P.M. on the day of the shooting. Six interviewers completed 312 three-question interviews.

Credibility Score. 23/35 = .66.

Generalizability 0. The study was of a single location, with no discussion of other studies of the spreading of news. (Compare this with Example 2.13.)

Discussion of Limitations 3. There is a discussion of the study's limitations–that it was done in New York City, that only persons home between 5:00 and 10:00 P.M. were interviewed, and that households with unlisted numbers were omitted. It would have been useful to know the completion rate, the number of refusals, the number not at home, and the number of men and women in the sample. Comparisons to other studies, such as the Kennedy assassination, also would have been valuable in interpreting these results.

Use of Special Populations 5. The use of a phone sample of the general population is appropriate here because of the need for speed. Random digit dialing (see Chapter 3) might have been possible.

Sample Size 5. About 60 interviews per hour were obtained. For the purposes of this study, this seems sufficient since the key question was merely "Have you heard the news. . .?".

Sample Execution 5. Although the details are sketchy, the brevity of the questionnaire suggests that very few of the persons who were reached refused to answer the three questions. Obtaining more than 300 interviews in an evening is an accomplishment.

Use of Resources 5. Although this obviously is not a perfect sample, it is a very good example of the optimum use of limited resources to collect timely data when a significant event has occurred.

Example 2.11 "Alienation and Action: A Study of Peace-Group Members" (8)

Peace group membership is related to various social–psychological variables, especially the mode of explaining social events. The sample was drawn from an unnamed suburban university community of 11,000 persons. The sample consisted of 30 "radical pacifists," chosen haphazardly from outside the community, 93 members of three peace groups (the total membership of these groups in the community), and 222 nonmembers, based on a sample of 10% of community households.

Credibility Score. 20/35 = .57.

Generalizability 0. The study deals with a single university community, with no comparison to other areas. Almost certainly, data for large cities with broader scopes of occupations would be different, not to mention the variation caused by regional and racial differences.

Discussion of Limitations 5. There is a very detailed discussion of the sample's limitations, pointing out that many of the members are university-related.

Use of Special Populations 5. Peace group members were used because of the topic being studied.

Sample Size 2. Beside the major problem of the sample's nonrepresentativeness, the sample is too small. The larger sample of the general population is of no value because the two groups of particular interest are the peace group members and the radical pacifists. Although all the peace group members in the single community are included, it would have been possible to increase the sample by adding other communities to the study.

Sample Execution 5. The 83% cooperation rate of the general population sample and the almost 100% cooperation of peace group members in the community indicates that the field work was carefully done.

Use of Resources 3. Too much of the resources in this study were used for the general public sample and not enough on the groups of prime interest. For comparison purposes, the optimum design would have been to have equal sample sizes of peace group members and the general public, if the collection costs were the same.

Example 2.12 "Ministerial Roles and Social Actionist Stance" (56)

This study reports the results of a factor analysis of Protestant clergymen in five cities to items on the role of the clergy. Two roles—traditional and community problem solving—are especially related to protest orientation. Of the 960 respondents selected in Atlanta, Boston, Los Angeles, Minneapolis, and Pittsburgh, 443 returned mail questionnaires.

Credibility Score. 19/35 = .54.

Generalizability 2. Although five cities are used here, all the results are combined, so that one has no notion of the variability between a Southern city, such as Atlanta, and the others, all with their own special characteristics.

Discussion of Limitations 2. While there is a general discussion of cooperation rates by location and denomination, there is no discussion of why the specific cities were chosen or how they might differ from other cities.

Use of Special Population 5. The use of Protestant clergy is a direct function of the purpose of the study.

Sample Size 5. The sample size of 443 is large enough for the analyses of this study but would be thin if there were separate analyses by denomination or city.

Sample Execution 2. The 46% cooperation rate could easily have been raised above 80% with two additional mail follow-ups. Although there are probably no education biases, there are obvious biases by city (cooperation ranged from 32% to 60%) and denomination (cooperation ranged from 40% for Baptists to 64% for Lutherans), and possibly by social action variables also.

Use of Resources 3. The use of mail is a good idea for this study, but there is no obvious reason for the study's having been limited to only five cities. Assuming directories were available for several hundred cities, a sample of 20 or more cities would not have required any additional mailing expenses, and would have required only a little extra for sampling. The results would have had far greater generalizability; also, some of the resources should have been retained for additional mailings.

Example 2.13 "Interpersonal Communication Following McGovern's Eagleton Decision" (62)

A street-corner survey in midtown Manhattan was conducted 4 days after McGovern's decision to drop Eagleton as his Vice-Presidential candidate. The sample of 108 respondents reported that interpersonal communication played little role in establishing awareness or changing attitudes, although they engaged in extensive discussion about the event.

Credibility Score. 18/35 = .51.

Generalizability 4. It is evident that the generalizability of this study is very limited since it was conducted only in midtown Manhattan. What saves it is the careful comparison of the results of this study to the results of 17 other studies of how persons become aware of news events (see Table 2.2). Other parts of this study, containing no comparisons, have lower credibility.

Discussion of Limitations 2. There is a very brief discussion of the limitations of the study, suggesting that 15% of passing pedestrians who were asked to cooperate refused; but there is no discussion of the possible biases of using pedestrians, or of what hours or days were used for interviewing.

Use of Special Populations 5. Aside from the biases, the respondents were part of the general population.

Sample Size 4. Although the sample size is small, the data are never split into more than three groups, so the sample size is barely large enough for the limited analysis.

Sample Execution 0. This is a completely haphazard sample. There is no real way of evaluating the quality of the sample or possible sample biases.

Table 2.2
Percentage of Respondents Who First Heard of Event through a Given Medium

	Communication media						
	Total broadcast	Radio	Television	Newspapers and magazines	Inter-personal	Cannot recall	Sample size
Kennedy's assassination							
Banta (Denver)	22%	13%	9%	0%	76%	2%	114
Burchard (DeKalb, Illinois)	35	22	13	0	65	0	100
Greenberg (San Jose)	47	26	21	0	53	0	419
Hill and Bonjean (Dallas)	43	17	26	0	57	0	212
Mendelsohn (Colorado)	56	39	17	0	32	12	200
Sheatsley and Feldman (National)	47	—	—	4	49	0	1384
Spitzer and Spitzer (Iowa City)	44	25	19	1	55	0	151
Average (Kennedy Studies)	42	24	18	0	55	2	
Studies from Deutschmann and Danielson							
Ike's decision to run after heart attack (Lansing, November 1957)	70	32	38	12	18	0	205
Explorer I Satellite (Lansing, January 1958)	60	20	40	17	23	0	167
Explorer I Satellite (Madison, January 1958)	65	29	36	22	13	0	125
Explorer I Satellite (Palo Alto, January 1958)	79	18	61	10	10	0	38
Alaskan Statehood							
(Lansing, June 1958)	52	32	20	33	15	0	84
(Madison, June 1958)	58	24	34	41	2	0	165
Studies from Budd, McLean, and Barnes (1964)							
Krushchev Ouster (Iowa City)	69	34	35	12	19	0	320
Walter Jenkins scandal (Iowa City)	47	25	22	50	2	0	230
Average (non-Kennedy studies)	63	27	36	25	13	0	
McGovern's Eagleton decision (New York)	74	34	40	11	15	0	108

Use of Resources 3. Although it might appear that street-corner sampling is an efficient use of resources, the biases from this method are so great that it is seldom used by current researchers who want their results to have any generalizability. Much better in this case would have been the use of telephone interviews, preferably with random digit dialing, as in Example 2.10.

Example 2.14 "A Test of Lindesmith's Theory of Addiction: The Frequency of Euphoria among Long-Term Addicts" (51)

Data are presented to show that long-term addicts experience euphoria frequently, crave it, and act to obtain it. The sample consists of 64 addicts in Baltimore—47 white and 17 black, 4 female and 60 male.

Credibility Score. 18/35 = .51.

Generalizability 0. The study is in only one city, with no comparisons to other studies.

Discussion of Limitations 4. A fairly good discussion of the characteristics of the sample is provided, but no discussion of how the sample might differ from the population of addicts. Thus, one suspects the sample is low on women, older addicts, and possibly blacks.

Use of Special Populations 5. The use of addicts is required by the theory being tested.

Sample Size 2. The data are adequate to prove that addicts experience euphoria, but the sample sizes are too small for most of the other analyses.

Sample Execution 2. This is really the most difficult aspect of the sample to criticize. On the one hand, it is evident that no list of addicts exists and that no screening procedure could produce a probability sample of addicts. On the other hand, this sample is clearly biased, in many ways, toward addicts most like the three male interviewers and most involved in drug-using social networks. It would be useful to add other kinds of users to the sample, such as those currently undergoing medical treatment and those in prison or awaiting trial. If the results were confirmed for these other groups, the study would be substantially more credible.

Use of Resources 5. The use of personal interviews with addicts seems to be the optimum procedure, although it would have been valuable to extend the sample to other locations by cooperating with other investigators, if this were possible.

Example 2.15 "Aspirations of Low Income Blacks and Whites: A Case of Reference Group Processes" (46)

A sample of 67 blacks and 110 whites in public housing in New York City are analyzed to determine whether their aspirations are a function of their position in their own racial group or of the position of the racial group in the society.

Credibility Score. 14/35 = .40.

Generalizability 0. This is a special sample of low-income households in only one location, with no discussion of results from other studies.

Discussion of Limitations 4. The study includes a rather complete discussion of the characteristics of this sample, without any discussion of population characteristics.

Use of Special Populations 2. The use of public housing residents as representing low-income households is primarily for convenience and not because it is necessary for the theory, as in Example 2.7. To find other low-income households would require screening the general population, a difficult task.

Sample Size 1. For many of the key analyses in this study, there are fewer than 20 respondents, and seldom ever are there more than 50 or 60. Thus, the detailed analysis is subject to very large sampling variability.

Sample Execution 5. The data for this study are a secondary analysis of data collected earlier by Caplovitz. As far as can be determined, the field work on the original study was done carefully.

Use of Resources 2. This study is an example of the secondary analysis of data initially intended for another purpose. The initial study was intended to prove that low-income households in New York pay more for furniture and other goods. The current study omitted about 60% of the initial sample because data for these respondents were not applicable or available. Although secondary analysis is very inexpensive, the small size and special character of the secondary analysis sample make broad generalizations very risky. One suspects that there are other sources of data for secondary analysis of low-income households that, if added to these results, could greatly increase generalizability and credibility.

Example 2.16 "Measuring Individual Modernity: A Near Myth" (5)

A sample of 156 white males in the "Uptown" neighborhood of Chicago was used to study individual modernity. The results indicate that modernity scales tend to predict scores on anomia, alienation, and socioeconomic status as well as predict other measures of modernity. Conversely, measures of anomia and alienation appear to predict modernity scores almost as well as do the modernity scales.

Credibility Score. 13/35 = .37.

Generalizability 3. This sample is in a single neighborhood in a single city. There are, however, comparisons to studies in Africa, Brazil, Mexico, and Turkey.

Discussion of Limitations 3. There is a brief discussion of the characteristics of "Uptown."

Use of Special Populations 0. The use of white married males is not a function of the theory but, rather, a convenience for the researchers. (The study had another purpose—the study of family-planning norms.) There is no reason, however, to believe that using this population introduces special biases.

Sample Size 5. All the analyses are based on the total sample, so the sample size is adequate.

Sample Execution 0. The sample of white married males is a haphazard sample, so nothing can be said about response rate or representativeness of the sample, even for the single neighborhood.

Use of Resources 2. Personal interviews were used, which seems an optimum procedure. The cost of conducting interviews in several Chicago neighborhoods and of using probability procedures would not have been substantially greater than the haphazard method actually employed.

Example 2.17 "Race, Sex, and Violence: A Laboratory Test of the Sexual Threat of the Black Male Hypothesis" (72)

Data from a laboratory experiment with 84 white male students at the University of California at Santa Barbara are used to support the psychoanalytic view of racism—that it is a function of the sexual threat of the black male.

Credibility Score. 7/35 = .20.

Generalizability 0. The study makes use of data for students at only one school, and there is no discussion of similar findings elsewhere.

Discussion of Limitations 2. The characteristics of the students participating in the experiment are given, but no comparison is made to the population of all college students or to the general population.

Use of Special Populations −5. Clearly, the college students are selected here for convenience, not to support a theory. Although the author believes that the theory is supported because the students are generally liberal and educated, my judgment is that this is just the sort of situation in which students will respond to the authority of the experimenter and give the expected results. Thus, I find these results totally unconvincing. The same results on a general population sample, even if limited geographically, would make a substantial difference in credibility for this kind of study.

Sample Size 2. The sample is split into four treatment groups for analysis, and because of an error in operation, one group has only 14 subjects while the other three groups have 27, 23, and 20 subjects.

Sample Execution 5. There is no evidence of bias in the class that was selected for this experiment—20 subjects were eliminated because they indicated an awareness of other similar experiments or because of an error in conducting the experiment.

Use of Resources 3. The ready availability of college students makes them the subjects for most experiments conducted by university researchers. Yet, as this example indicates, this procedure may seriously affect the credibility of results. The use of noncollege populations instead of, or in addition to, college students can greatly improve the quality of a sample.

It should be obvious to readers that there are very substantial differences in the quality of the samples presented in these examples; those given first are clearly better than those discribed later. There seems to be no value in arbitrarily assigning words like "good" and "poor" to the studies at the top and bottom of the list, since the changes in quality are continuous rather than discrete. As anyone who has ever graded examinations knows, it is far easier to recognize high quality than to decide how much poor quality should be penalized.

It should also be evident from the examples that limited resources need not necessarily lead to low-quality samples. The imaginative use of special populations when applicable, collaboration with other researchers, and comparisons with other studies all help to improve the quality of a sample. The appropriate use of mail and phone methods should always be considered. As is stressed throughout the remainder of this book, some careful thinking early is always better than later regrets and apologies.

2.4 SUMMARY

In this chapter are described a group of studies, none of which use standard national probability samples. There are, however, major differences in the quality of these studies, based on the following criteria:

1. How well can the data be generalized?
2. How complete is the discussion of the sample and its limitations?
3. Is the use of a special population necessary to test a theory or is a special population used only for convenience?
4. Is the sample size adequate for the analysis reported?
5. How well was the sample design carried out?
6. How well were the limited resources used?

Several procedures are suggested for maximizing limited resources—working with colleagues at other geographic locations, comparing results to other published work, and using special populations and mail and phone methods where applicable. The discussion of sample limitations should include not only a description of the sample but of the population that the sample is intended to represent.

Small samples, of necessity, limit the depth of analysis. The solution is not to ignore these limits in discussing the data but, rather, to oversample critical groups that otherwise would be underrepresented. There should always be a description of how the study was conducted, the methods used, and the cooperation rate achieved. These results are, of course, meaningful only if a careful design and not a haphazard sample is used.

Ingenuity and careful planning can overcome many of the problems associated with limited resources and can produce samples with high credibility.

2.5 ADDITIONAL READING

The most useful additional reading would be some or all of the studies cited in the examples of this chapter, with a critical eye on the sampling methods. Readers may wish to compute their own credibility scores and see how these scores agree with or differ from the ones given in this chapter.

The next step would be to apply the same criteria to other articles found in professional journals of special interest, or in monographs reporting sample results. Many of the references at the end of Chapter 1 would also be useful, if not consulted previously.

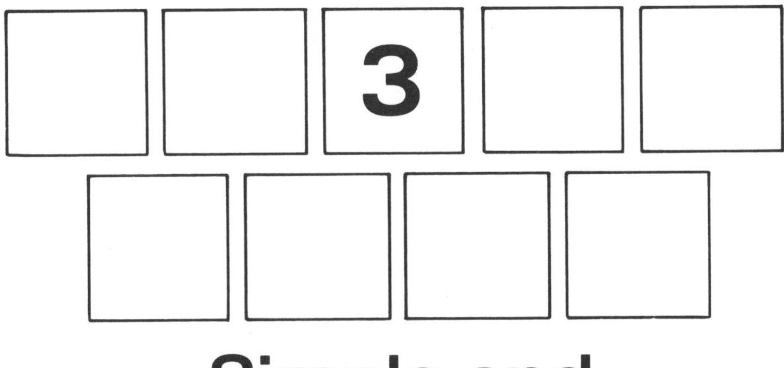

3 Simple and Pseudo-Simple Random Sampling

A researcher with limited resources is indeed fortunate if the population he is studying and his data collection methods allow him to use simple random sampling procedures. These procedures are easy and inexpensive for sample selection, data analysis, and sampling variance computation.

It is important to recognize, however, that, in many cases, simple random sampling is not appropriate. We shall discuss first the general conditions when simple random sampling is optimum, and give some illustrations. In contrast, we shall then give examples of problems for which simple random samples are not appropriate, either because there is no way to select a simple random sample or because other more complex methods are also much more efficient. When other methods discussed in later chapters are preferable, it would be a serious mistake to use simple random sampling procedures merely because they are easy.

Probability samples are defined as those samples in which every element in the population has a known, nonzero probability of selection. These samples are sometimes called *random samples,* but since the term "random" is used in different ways by statisticians and samplers, the term "probability sample" is less likely to cause confusion. Note that it is not necessary that the probabilities of selection be equal for all elements, and this will not be the case for the more complex samples discussed later. *Simple random samples,* however, are defined as those in which (*1*) the probabilities of selection are equal for all elements, and (*2*) sampling is done in one stage with elements of the sample selected independently of one another, in contrast to more complex samples in which the selection is done in two or more stages and in which clusters rather than individual elements are chosen.

3.1 RANDOM NUMBERS

The word "random," as used in sampling, does not mean haphazard or "catch as catch can" but, rather, implies that some well-designed probability mechanism is used in the sample selection. It is the use of this probability mechanism at each point in the selection that distinguishes probability samples from judgment samples.

When one first thinks of ways of obtaining a random process, one might consider the use of playing cards, dice, spinners or roulette wheels, or drawing names or numbers out of a hat or a fishbowl. The problem with all these procedures is that they are slow and that there are major difficulties in maintaining randomness in the long run. Thus, playing cards must be shuffled, and imperfections in the shuffling lead to the repetition of sequences already observed. Expert card players make use of nonrandomness in shuffling by remembering how cards were played in the previous deal.

A classic example of the difficulty of obtaining randomness by using numbers in a fishbowl was the initial use of this procedure for determining the order of selection of men for the draft during the war in Viet Nam–using birthdates. The dates of the year were poured into a bowl, January first and December last. Although the bowl was then stirred vigorously to obtain randomness, a much higher number of December dates were chosen early while January dates were chosen later. Subsequently, the procedure was revised so that there were two bowls. A date of the year was selected from the first bowl and an order rank from 1 to 365 was selected from the second. Much greater care was also taken in the way both bowls were filled and mixed, and the new process appeared to be essentially random.

Even dice and roulette wheels are subject to uneven wear after hundreds of thousands of uses and become nonrandom. The most convenient and accurate procedure for obtaining a random process is through the use of tables of random numbers. The largest printed table is *A Million Random Digits* by the Rand Corporation (67). Four pages from this table are given in Appendix A, pp. 223–226. The table is also available on IBM cards for use in computer programs.

The Rand random digits were generated by an electronic roulette wheel. A random frequency pulse source passed through a 5-place binary counter and was then converted to a decimal number. The process continued for 2 months and, even with careful tuning of the equipment, the numbers produced at the end of the period began to show signs of nonrandomness, indicating that the machine was running down. This nonrandomness was eliminated by adding together pairs of digits. This illustrates the difficulties in attempting to generate random numbers by virtually any process.

To know if the numbers in a table of random numbers are indeed random, several statistical tests can be performed. The primary test is to observe the distribution of the digits 0 through 9 for the total table and parts of it. One expects that each digit will occur 10% of the time. Overall, in the Rand table, the frequencies vary from 9.93% for the digit 9, to 10.06% for the digit 2, indicating no significant deviation from randomness. Other tests include "poker" tests of 5 digits to count the duplication of digits in blocks of 5 and runs tests to look for evidence of serial correlation between numbers.

How does one use a table of random numbers for sampling? One of the first questions that worries a new user is where to begin. The simple answer is that it does not matter. One may begin anywhere in the table and move in any direction. An easy procedure is to start at the beginning and read down. The only error to avoid is starting at a given place because one knows the distribution of numbers at that place. An even worse sin is to reject a sample because somehow it does not look right, and to continue using random numbers until a likely-looking sample is selected. Obviously, this destroys the sample's probability character and makes it a judgment sample. (In some cases, however, the initial sample selection may uncover serious problems, if simple random sampling is used, and suggest that a more complex procedure is required. A new sample draw, using optimum procedures, would not be cheating on probability methods.)

Once a starting point is selected, the number of columns of digits read must be sufficient to give each element in the population a chance of selection. If there are 70,000 elements in the population, it would be necessary to use 5 columns of digits; if 900 elements in the population, only three columns would be needed. The groupings in the table itself are only for ease in reading.

To make the random selections, the elements of the universe are numbered, and the selection of random numbers continues until the desired sample size is reached. Some of the random numbers chosen will be larger than the size of the universe and these numbers are discarded, as are any duplicate numbers. Thus, in sampling the population with 70,000 elements, any number starting with 7, 8, or 9 would be ignored except for the number 70,000 itself.

If one is selecting from a population of which the size of the first digit is small, such as sizes of 250, 3400, 11,000 there will be many more random numbers discarded than used. Some statisticians have suggested procedures for using a higher fraction of these numbers, but the arithmetic, although extremely simple, becomes time-consuming if large samples are required. The easiest thing to do is to have a large table and to use only eligible numbers.

If the same random number is selected more than once, the second and subsequent selections are discarded. This means that the procedure is simple random sampling without replacement. The effect of this is to reduce the sampling error of the estimate, and the reduction is largest when the sample chosen is a substantial fraction of the total population. This is most easily seen when the total population is studied. Then there is, of course, *no* sampling error. The computation of sampling errors for sampling without replacement is discussed in Chapter 8.

Example 3.1 Sampling from the University of Illinois Student Directory

The University of Illinois at Urbana-Champaign Student Directory for 1972–1973 consists of a total of 33,271 names. The names are not numbered, but, since the book is set from an IBM printout, there are exactly 97 names on each of the pages. This fact is useful, since it eliminates the need to count some pages and provides a check for manual counting. Alternatively, it would be possible to obtain a new printout already numbered.

Suppose one wishes to select a sample of 20 students for a pilot test of a new questionnaire. Using the abbreviated table of random numbers (Table 3.1), one would read the numbers in groups of five to give each student a chance of selection. Reading first down and then across, the first five numbers are ignored because they are too large. The next four numbers are chosen, and then three numbers are skipped, and so on. The process continues until the twentieth number is selected; then the sampling stops.

3.2 SYSTEMATIC SAMPLING

The use of simple random sampling may be a long and tedious process if both the sample and population are large and manual procedures are used. Suppose, for example, that one has a list of 1 million inhabitants or voters in a large city and wishes to choose a sample of 1000. Intuitively, most people, when asked how to sample, would reply, "Take every thousandth case." This procedure of taking every ith case is widely used by professional samplers and is called *systematic sampling*. Because of its simplicity and usefulness in complex sampling situations, systematic sampling is probably used far more frequently than simple random sampling.

To do systematic sampling, just two things are needed—the *sampling interval* and a *random start*. The sampling interval, if one has a list and wishes to approximate simple random sampling, is merely the ratio $i=N/n$ of the number of elements in the population (N) to the desired sample size (n). In selecting more complex samples, as discussed in Chapter 7, the sampling interval is computed in a slightly different way, using measures of size.

A table of random numbers is used to select the initial number between 1 and i, called the "random start." This ensures that every element in the population

Table 3.1
Table of Random Digits

78986	45691	79922	40294	52672	46262	58177	55586
83230	59025	72573	(10) 18282	45513	82933	(17) 27817	47485
58846	(5) 01946	(7) 00367	38926	58508	36119	(18) 20874	35592
51999	(6) 19130	90645	68287	33553	38330	(19) 13265	99744
61096	59042	57643	(11) 00032	79958	(14) 08614	71178	23270
(1) 30226	69093	63119	84323	(13) 28281	49514	(20) 26440	24786
(2) 02073	65554	56777	79666	40379	(15) 12544	24225	63822
(3) 05250	86448	68145	82707	79180	95248	74151	48197
(4) 08014	95229	(8) 03319	(12) 30045	59371	95039	13334	14496
76489	52722	(9) 25901	44752	99943	(16) 08909	16161	92356

has an equal chance of selection and guards against the possible small bias that might occur if the first or last member in the interval always were selected. In the sample of 1000 from the list of 1 million residents, suppose the random start selected is 243. Then, the selected elements in the population would be 243, 1243, 2243, 3243, . . . 999,243.

USE OF LENGTH MEASURES

Although systematic sampling eliminates the extensive use of a table of random numbers, it still would appear to require complete counting of the universe. If the sampling is done manually, the use of some shortcuts can reduce this counting procedure considerably. Consider a city directory that has two columns per page, 100 names per column, and 500 pages—or a total of 100,000 names, all unnumbered. If a sample of 1000 names is required, it is easy to see that this can be obtained by selecting 1 name per column. Once a random start is selected and counted in one column, the distance of the random start from the top of the page may be measured on a strip of cardboard and this strip of cardboard placed on subsequent columns to locate the sample members.

Usually, however, the column length will not be identical to the required interval. If the column length is shorter than the interval, this would mean that selecting one unit per column would give a sample larger than required. In this case, it might be easier to choose the larger sample and then to delete systematically until the proper sample size is obtained. *A useful property of simple random samples is that a simple random subsample of a simple random sample is also a simple random sample.*

If, as is often the case, the sampling interval i is not a whole number, the easiest solution is to use as the interval the whole number just below or above i. Usually, this will result in a selected sample that is only slightly larger or smaller than the initial sample required, and this new sample size will have no noticeable effect on either the accuracy of the results or the budget. For samples in which the interval i is small (generally for i less than 10), so that rounding has too great an effect on the sample size, it is possible to add or delete the extra cases. As already suggested, it is usually easier to round down in computing i so that the sample is larger, and then to delete systematically.

Example 3.2 Systematic Sample from the University of Illinois Student Directory

Using the same student directory as in Example 3.1, let us select a systematic sample of 300 students. The sampling interval is 33,271/300 = 110.90, or 111. Since there are 343 complete pages of 97 names and 1 partial page, a systematic sample of 1 name per page will yield a sample of either 343 or 344 names depending on the random start selected. Then, to obtain a sample of 300, note that 300/344 = .87, or just about 7/8. Thus, deleting every eighth name after a random start yields a sample of 300 or 301. This procedure may seem more complicated, but in practice it is far faster than having to count intervals of 111 cases. An equivalent procedure would be first to randomly delete one-eighth of the pages in the directory and then to select 1 name per page from the remaining pages.

Another method, if the interval is longer than the column length, is to measure the length of the interval on a cardboard strip, being careful to omit margins, and then to measure off intervals on the cardboard strip. This is subject to possible errors in placing the strip but, if done carefully, should have no noticeable bias on the results.

The use of measurement methods may be very helpful if the population is not printed on a list but is arranged on cards or in file drawers. For the sample to be unbiased, however, it is necessary that all files be of the same thickness, otherwise a thicker file has a higher probability of being selected. It is also necessary to ensure by some spot checking that the density of files per unit of length is reasonably identical throughout the files.

If the sampling interval is smaller than the number of units per page, several selections per page may be made at random and the cardboard strip marked to indicate each of the selections.

Example 3.2 (Continued) Systematic Sample from the University of Illinois Student Directory

Continuing our example of systematic sampling from the University of Illinois Student Directory, suppose one needed to select a sample of 1000 students. The sampling interval is now 33,271/1000, or 33.3. Although it would be possible to take every thirty-third or every thirty-fourth student, this would require counting and numbering all students in the directory. Noting that there are 97 names per page, if one divides 97 by either 33 or 34, one would get slightly less than 3 names per page (2.9). Rather than counting, it is far easier to make 3 selections per page and then to use a cardboard strip to locate sample members. If 3 selections per page were made, this would yield a sample of 1029 to 1032 names, depending on how many are selected from the partial final page. Randomly deleting 10 pages with 3 names each, or every thirty-fifth name, would yield the approximate desired sample; or one might decide to keep the larger sample.

All these procedures may be programmed and executed on a computer. If the population is already on tape or IBM cards, the computer run will be far more efficient than these manual methods suggested. In this case, fractional sampling intervals need not be rounded to whole numbers, since the computer can handle fractions with no loss of efficiency. If the data are in printed form or in files and a complete computer tape is not required except for sampling, it will usually not pay to use the computer, because the manual sampling will be faster than the data preparation and punching for the computer.

3.3 ARE SYSTEMATIC SAMPLES SIMPLE RANDOM SAMPLES?

The reader may wonder why this discussion of systematic random samples is in the chapter on simple random samples. Are the two kinds the same? The theoretical answer is that systematic samples are really complex samples with unknown properties and that they may be substantially different—sometimes better, sometimes worse—than simple random samples. A very careful discussion of theoretical possibilities is given by Cochran (14). The practical answer is that, in those cases for which simple random sampling is appropriate, simple random samples and systematic samples will be about the same except in very unusual situations of periodicities. When using systematic sampling, it is important to inspect the lists before beginning in order to ensure that there are no obvious periodicities. Nevertheless, in more than 20 years of experience, I found only one case in which a systematic sample produced very strange results, and this was evident as soon as the sample was selected.

Example 3.3 Periodicity in a Systematic Sample

Residents in several communities were interviewed about their attitudes toward the neighborhood and their satisfaction with neighborhood services. The sampling for this study was a complex multistage sample of the types discussed in Chapter 7. Three years later, a follow-up study was done with a subsample of one-eighth of the initial sample, to study changes in attitudes. This sample was selected by using systematic sampling with a sampling interval of eight and a random start of four. Surprisingly, although the initial sample had been about half male and half female, this systematic sample was all female.

An investigation revealed that initial respondents had not been numbered sequentially as the questionnaires were returned, as is usually done. Instead, the final digit of the respondent number was used to indicate sex of respondent, even numbers representing women and odd numbers men. This numbering scheme was not known until the sample was selected, but since both the random start and sampling interval were even, the sample consisted entirely of women. Once this was discovered, the sample was discarded and another selected after first sorting out the sexes.

This example illustrates one of the possibilities for making a systematic sample worse than a simple random sample—periodicities in the list. Other periodicities are conceivable, such as in lists of military personnel arranged by platoons, factory workers arranged by work units, or elementary school students in classes. However, in the real world, it is very unlikely that there would be exact periodicities corresponding to the sampling interval when dealing with lists of people.

One case in which one should avoid systematic sampling is the sampling of time periods. Suppose one were going to sample traffic on a street or shoppers at a store that was open 24 hours a day. With any sampling interval, it is difficult to avoid some time periodicities. For example, if the decision is made to sample 12 quarter-hours during the week, the sampling interval would be 56, since there are 672 quarter-hours per week. In this case, the sample would consist entirely of all odd or all even hours, and the same quarter-hour segment for every hour selected. (The reader should verify this by choosing a random start and listing the selected times.)

In the sampling of nonhuman populations, such as businesses, financial accounts, temperature or climate, and lumber or crop yields, there may be not only periodicities in the data but also some linear trends. Thus, savings deposits measured over time may show substantial increases due to inflation. For these cases, neither simple random sampling nor systematic sampling is appropriate. The optimum sampling procedures involve the stratified sampling discussed in Chapter 6.

The claim has sometimes been made for systematic sampling that it is more efficient than simple random sampling because it eliminates autocorrelation in the sample, that is, the similarity of adjacent elements. On a voter list, a systematic sample would not select members from the same household, while this could occur in a simple random sample. In practice, this improvement is so small that it can be ignored. If autocorrelation is an important issue, as in economic time series, then it is again better to use the explicit stratification methods of Chapter 6.

There is an important exception where systematic sampling may substantially improve sample efficiency by being a form of implicit stratification. This is the case when the sampling is of large geographic areas, such as counties or other clusters, rather than of individuals. The detailed discussion of this situation is given in Chapter 7, which discusses multistage samples.

To summarize, systematic samples are not really simple random samples, but they behave as such and have the same precision in almost all cases of interest involving human populations. They are used instead of simple random samples, not because they produce better data but because they are easier to use. Systematic samples are often described by samplers as "pseudo-simple random samples."

3.4 THE USES AND LIMITATIONS OF LISTS

Probably the most difficult task in sampling from lists is finding the appropriate list. It may be helpful to mention the kinds of populations for which no lists exist. There is no list of all the population or households in the United States. One might consider using the Census of Population and Housing, but this information is closely guarded by the Census Bureau to protect confidentiality and is never released for sampling purposes. The Census Bureau will consider requests to select samples if all the sampling is to be done by Census Bureau personnel and no confidential information is released. Most readers of this book, however, who might need a national population sample would be better off using one of the existing field organizations with a national sample, as discussed in Chapter 1.

From the discussion here, it follows that there are no population lists of men or women, young people, old people, blacks, or whites, nor lists categorized by income. A researcher sometimes hears of mailing lists that provide this sort of information, but, though mailing lists may be useful for people trying to sell something, they are usually worthless for sampling. Mailing lists are derived primarily from membership and subscription lists, and it is not possible to specify the population they represent. Thus, a sample selected from a mailing list of boys under 16 would probably be based on membership in the Boy Scouts or on subscription to *Boy's Life* magazine, and would obviously not be a sample of all boys.

At the local level, however, population and household lists are available for most medium-sized cities in the range of about 50,000 to 800,000 people. About 1400 of these directories are published by R.L. Polk and Company (6400 Monroe Boulevard, Taylor, Michigan 48180). The directories are usually available for free use at the local public library or the chamber of commerce office, as well as for sale by Polk. For sampling purposes, most directories contain both an alphabetical list of names of residents and businesses and a street address directory of households. Since the directories are revised every 2 or 3 years, the street address directory is reasonably accurate. It misses only new construction that occurs after the directory is published. The alphabetical list is subject to greater error over time, for many families and individuals will move in or out of the area or to some new address. Although the directories are subject to listing errors, these are usually corrected in subsequent directories. Overall, the quality of the lists is usually as good or better than the lists that could be obtained by a careful researcher who starts from scratch with new listers. Of course, the use of existing lists is far cheaper.

If the population consists of members of some professional organization, there will probably be a membership directory published. Some organizations, such as the American Medical Association and the American Dental Association, have put their directories on tape for mailing and sampling purposes. If one does not know the complete name and address of the organization, the easiest place to find this information is in the *Encyclopedia of Associations* published by the Gale Publishing Company of Detroit.

Lists of business establishments grouped by SIC (Standard Industrial Classification) codes may be purchased from Dun and Bradstreet, New York City; these lists are revised relatively infrequently, so they do contain names of establishments that are no longer in business. Lists of schools and colleges are available from publications of the U.S. Office of Education and from state and local superintendents of education. Ordinarily, lists of teachers and students are not available for elementary and high schools but are available in published directories for individual universities. National mailing lists of college students and teachers have been prepared, but they should be used cautiously because, in the past, schools have been omitted and the lists quickly become outdated.

3.5 BLANKS AND INELIGIBLES ON LISTS

There are three common problems in the use of lists. The treatment of these problems is discussed in this and the next two sections. The first problem is that the list may contain blank or ineligible units. Suppose one uses the systematic sampling procedure discussed earlier, in which the selected unit on a page is obtained by measuring down from the top of the page. For explicitness, assume the sample is to be selected from a city directory with the names in alphabetical order. The population to be studied may consist of either households or individuals. The selected line on the page, however, instead of containing a name of an individual, may list the name of a business firm or professional office or, in some listings, a school, park, or government office. These are ineligible listings. Blanks are obtained if there is no listing at all on the line, due to spacing between letters of the alphabet or because the column ends in the middle of a page, or if a name is on the line but the address given is incorrect because the unit has been demolished or is vacant.

One might first consider going through the list and removing all ineligible listings, but, in a large list like a city directory, this is clearly impractical.

Two intuitive procedures for solving the problem of blanks and ineligibles usually occur to the naive sampler. Unfortunately, both of these naive procedures introduce serious biases into the sampling. The first *incorrect* procedure is to take the next name on the list; the other *incorrect* procedure is to count down a fixed number of eligible names from the top of the page. Both these methods give a higher probability of selection to people or households whose names follow ineligible listings. In some cases, it is hard to know what kind of bias this introduces into the results, but, in directory listings, it is likely that the name after a business or professional listing will be the home address of the same person. Thus, using the incorrect procedures yields a sample with too many doctors and butchers and too few college professors and other working types.

The correct solution to the problem of blanks or ineligibles is really straightforward. They should be ignored—excluded from the sample. This requires, however, that the sampling interval be adjusted to account for the expected number of exclusions. If one knows that $p\%$ of some list is eligible, while $(1-p)\%$ consists of ineligibles or blanks, the sampling interval is $i=Np/n$ where, as before, N is the total size of the list, Np is the number eligible, and n is the desired sample size.

Estimation of the proportion of the list that is unusable can be done on the basis of prior experience with the list or by counting the proportion ineligible on a sample of 5 to 10 pages. If this estimate is either too high or too low, additional sampling may be required to obtain the desired sample size, or it may be necessary to make some random deletions of the selected sample.

While considering the problem caused by ineligible or blank listings, samplers generally make an estimate of the cooperation rate likely to be obtained from the selected sample. In national samples, experience indicates that most field organizations obtain about 80% cooperation from the designated sample. Interviewers from the Census Bureau obtain higher cooperation, about 95% on the monthly Current Population Survey, and very high cooperation rates of 95–100% are obtained by many organizations in small towns and on farms. As city size increases, the cooperation rate drops to as low as 70% in the largest metropolitan areas. On the basis of past experience with similar populations, one should make a realistic estimate of the cooperation rate c and oversample to obtain the desired sample size. The sampling interval i is then computed by $i=Npc/n$.

Example 3.4 Computing the Sampling Interval *i* for a City Directory

Suppose one wishes to select a sample of 1000 completed cases from the city directory for Peoria, Illinois. The directory has 367 pages with three columns per page and 84 lines per column, or a total of 92,484 lines. Note that it is not necessary to assume, or is it the case, that each column has the same number of households. Some columns will have more businesses or blank lines. If one of these lines is selected, the line is not used. A systematic sample of 10 pages indicates that 63.1% of the lines contain a household listing, 9.0% contain a business listing, 17.2% contain a street name or Zip Code, 7.4% are blank, and 3.3% contain advertising. Assuming that the cooperation rate is expected to be 80%, the sampling interval would be 92,484 (.631) (.80)/1000, or 46.7.

The easiest procedure would be to note that selecting two lines at random from each column would produce a sampling interval of 84/2, or 42, which is smaller than 46.7 and would yield a larger sample. The expected sample based on these estimates is about 1389, although this is only an estimate. An actual sample was selected by using this interval and two random lines per column, numbers 5 and 27. The number of households selected was 1441. Since only 1250 are needed, assuming a cooperation rate of 80%, one computes the ratio of 1250/1441 and notes that .867 is approximately equal to the fraction 7/8, which is .875. Thus, randomly deleting every eighth element in the sample of 1441 after a random start yields a sample of 1260. Usually at this stage, the sampling would end: Given the uncertainty about the final cooperation rate, the deletion of 10 additional cases to reach exactly 1250 would not be very important.

3.6 DUPLICATIONS

In simple random sampling, each element of the population must have an equal probability of selection. If some individuals or households appear on the list more than once, some possible biases are introduced. Although often these biases will be very small, they sometimes may be serious.

Example 3.5 Duplicate Entries in a Contest

A candy company once ran a contest and received several hundred thousand entries. The marketing staff were interested in estimating the average number of entries submitted by persons who had entered the contest. A sample of entries were selected and a count was made of the number of times entries had been submitted by the persons submitting the sample entries. The company was upset to learn that, based on their sample, the average person submitted more than 50 entries. An outside researcher who was informed of these results pointed out, however, that the sample was heavily biased toward persons making multiple entries. The probabilities of selection for the sample were not equal but were proportional to the number of entries. The correct estimate was actually much smaller, only about 7 entries per person.

To make the discussion more explicit, suppose there is a contest in which 330,000 entries are received from 100,000 different persons. The distribution of the number of entries is given in Table 3.2. It is obvious that the actual mean number of entries is 3.3 per person. If a 1 in 1000 sample of entries is selected, this sample will heavily overrepresent persons with a large number of entries, as may be seen in the last column of Table 3.2. If, mistakenly, a mean is computed from this sample of entries, the observed value will be 9.5, or about three times the actual mean.

Example 3.6 Duplications Using Multiple Lists

The American Public Health Association (APHA) wanted to select a sample of persons engaged in public health activities. No single list of such persons existed, but the combined membership lists of five organizations were thought to include

Table 3.2
Persons and Entries for Hypothetical Contest Example

Number of entries	Number of persons	Total entries	Sample of 1 in 1000 entries
1	50,000	50,000	50
2	20,000	40,000	40
3	10,000	30,000	30
5	8,000	40,000	40
10	7,000	70,000	70
20	5,000	100,000	100
Totals	100,000	330,000	330

$$\text{Actual mean} = \frac{330{,}000}{100{,}000} = 3.3$$

$$\text{Computed mean} = \frac{50 + 2(40) + 3(30) + 5(40) + 10(70) + 20(100)}{330} = \frac{3120}{330} = 9.5$$

virtually everyone in the field. It was recognized, however, that some persons might belong to two, three, four, or even all five organizations. Unless these duplicates were removed, persons more active in organizational activities as indicated by multiple memberships would have a higher probability of selection. In this case, since the APHA wanted an unduplicated mailing list for purposes other than the sample survey, the 50,000 total names were cross-checked and a net list of 37,000 names was obtained. The cross-checking of names is a costly task and need not be done for the entire list unless an unduplicated list is required for other purposes.

Example 3.7 A Sample of Students and Staff at the University of Illinois

A survey of housing needs of both students and staff at the University of Illinois was conducted, using both the student and the staff directories. Duplications arose because some students (particularly graduate students) are listed in both directories. A total sample of 1000 was desired. Using an interval of every forty-fourth name in both directories, a sample of 1049 were selected of which 96 were found on both lists. Half of these 96, or 48, were randomly omitted, leaving a sample of 1001 individuals all of whom had an equal probability of selection.

Some lists are arranged in ways that make it very difficult or impossible to determine duplications in advance. Thus, a register of manufacturing firms arranged by types of products might well list the same firm several times, and discovering the duplication would be difficult because there might be hundreds of product categories. Similarly, one could not determine multiple car ownership of a sample of car owners chosen from a list of current license numbers if the numbers were arranged in numerical sequence rather than alphabetically. In this case, it is possible to determine the duplication by asking a question during the survey interview to determine the number of times the person or firm was listed. ("How many cars do you own?" or "How many different products does this firm make?")

To keep the sample unbiased, it then would be necessary to discard some completed interviews so that all respondents would have equal probabilities of selection. It is wasteful to discard interviews that have already been collected. The better alternative would be to weight the results by the inverses of the probabilities of selection. Thus, sample elements that were discovered to have been listed k times would be weighted by $1/k$. This process is no longer a simple random sample, but is really a stratified one. Other examples of weighted samples are discussed in Chapter 5.

3.7 OMISSIONS FROM LISTS

The most serious problem in using lists is the omission of an important fraction of the total population. Small listing errors can usually be ignored, such as typographical and clerical errors in a city directory, but some lists, by their coverage, exclude important segments of the total population. A telephone directory for a city excludes persons without telephones and with unlisted numbers. A list of registered voters excludes recent movers and noncitizens as well as those who have chosen not to register. A list of the largest firms in an industry excludes all firms below the cutoff.

Three alternatives are possible for handling omissions in lists:

1. Discarding the list
2. Using the list and ignoring omissions
3. Using the list with supplementation

Each of these strategies is appropriate under certain conditions. Before deciding on the strategy to be used, the purposes of the study and the quality of the list should be considered. The quality of a list may be estimated by comparison to census data if available, or by comparison to other lists or new listings. Thus, if a voter registration list for a small town contained 2210 names and the 1970 census of population indicated a population of 2290 persons 18 and over, the voter list would be adequate and the few omissions could be ignored. If, on the other hand, a voter list for a large city contained only 60% of the population when compared to census figures, it probably would be discarded.

If there were uncertainty about the quality of a city directory, listers could be sent out to relist a sample of 20 blocks. If the new listing uncovered less than 2% or 3% errors, the list could be used with confidence for even a careful study. Any new list that a researcher developed would probably have at least 5% errors, according to some careful experiments conducted by Kish and Hess, as reported by Kish (40, p. 531). For preliminary studies or cases in which precise estimates are not needed, lists containing 80–90% of the total population are often used with omissions ignored.

Frequently the best way to obtain a careful sample while keeping costs of sampling low is to use an available list and supplement it with other procedures. This is analogous to the use of combined methods of data gathering discussed in Chapter 1, and takes us from the realm of simple random samples into stratified sampling discussed in detail in Chapter 6. Here we merely sketch a common procedure.

For many of the largest metropolitan areas, telephone directories list between 70% and 90% of all households. Careful studies cannot ignore, however, households without phones or with unlisted phones. To supplement the directories, listers are sent out to a sample of blocks to list households without telephones. (Methods for choosing these sample blocks are discussed in Chapter 7.) The lister is provided with a list of households which have listed phones, obtained from a street address phone directory. Such a directory is published for the 35 largest standard metropolitan statistical areas and lists phone households in geographic rather than alphabetic order. The lister then adds all households in the sample areas that are not already on the list.

There is a simple criterion for determining if combined methods are preferable to a totally new listing—the cost of the combined methods should be lower than the cost of the new listing. If there are no savings, the incomplete list is discarded. Usually, for studies of the general population, the difficulties in obtaining current voting lists and supplementing them with new listings make this method more costly than totally new listings, so voting lists are valuable only when one wishes to study voting behavior. Similarly, in studying low-income households, the combined use of welfare records and new listings is usually so difficult to integrate that new listings are commonly used.

3.8 THE USE OF TELEPHONE DIRECTORIES AND RANDOM DIGIT DIALING

Since so many researchers with limited budgets are using phone interviews, a brief discussion of methods for improving the quality of telephone directory listings may be helpful. Obviously, no method can affect those households without phones. If this bias is important, some face-to-face interviewing is necessary. It is possible, however, to obtain interviews from households with unlisted phones through the use of random digit dialing methods.

In their simplest form, random digit dialing procedures ignore telephone directories entirely and select numbers to be called by using tables of random numbers. Every working telephone number in the population has an equal probability of selection when this method is used. The problem with the procedure is that it also produces a very large majority of numbers that are not in use at all or that are used by businesses or other institutions. It is generally possible to obtain from local telephone companies the central office exchanges that are in operation–that is, the first three digits of the telephone number after the area code. It is not usually possible to get the numbers within a central office that are being used. Glasser and Metzger (31) estimated from a sample of 30,000 numbers that only 1 of 5 was usable.

A simple combination of directory sampling and random digit dialing is far more efficient than pure random digit dialing. One first selects a random (or systematic) sample of household listings from the directory and ignores the last three digits. A table of random numbers is then used to select these digits. It can be shown that this procedure greatly increases the probability of obtaining a working household number: With this procedure, about half of all selected numbers are usable.

There is some arbitrariness in the decision to ignore the last three digits, for it is possible also to ignore only the last digit or the last two digits of the selected number from the directory. This is most efficient in eliminating nonworking and nonhousehold numbers, but it increases the possibility of bias due to the missing of a new series of numbers being activated. Using all but the last three digits does not eliminate this bias entirely but does ensure that it will not be very large.

Before using random digit dialing, one should first estimate the proportion of unlisted phones in the population. Outside the very largest cities, the proportion of unlisted phones is under 5%, and the additional cost of two calls for one working number, using even the most efficient procedures, is probably not worth the small reduction in bias, remembering that households without phones are still excluded. In the largest cities and some of their suburbs, as many as 40% of all phones may be unlisted. It is in these places that random digit dialing is most appropriate.

If information on the percentage of unlisted phones cannot be obtained from the local phone company, an estimate can be made by comparing the number of household listings in the 1970 directory to the estimate given in the census of housing for the same place.

Example 3.8 Selecting a Sample from the Chicago Phone Directory Using Random Digit Dialing

The initial step in this process is the selection of listings from the directory using the systematic sampling methods discussed earlier in the chapter. Table 3.3 gives a small sample of five listings and the random numbers chosen from these listings. Note that, if the initial listing is not a household phone, no random

Table 3.3
Chicago Phone Directory Listings
and Random Digit Replacements

Initial listing	Type of unit	Random digits selected	Listing actually contacted
FI6–5656	commercial	–	–
276–*7915*	residential	064	276–*7064*
287–6871	commercial	–	–
HE4–*3501*	residential	392	HE4–*3392*
729–1*224*	residential	898	729–1*898*

substitution is made. This eliminates calls to exchanges that are used entirely for commercial purposes.

3.9 SUMMARY

The major focus in this chapter is on the use of tables of random numbers and of lists for selecting simple random samples. Shortcut procedures, using systematic sampling, are described as generally equivalent to simple random samples and as much easier to use. The major problems with lists are described along with their cures. Blanks on lists are simply ignored, but a smaller sampling interval is used to give the desired sample. Multiple listings are usually treated after the sample is selected by subsampling $1/k$ of the units that are listed k times.

If omissions are a serious problem with a list, supplementary listings are used in combination with the initial list. If this method is too costly and complicated, the initial list may be discarded entirely. Finally, the use of random digit dialing in combination with telephone directories is described for large cities where a substantial fraction of phones are unlisted.

3.10 ADDITIONAL READING

Random Numbers. For readers interested in knowing more about how random numbers are generated and tested, the introduction to the Rand Corporation's *A Million Random Digits* (67) is probably the best brief source.

Systematic Sampling. The discussions of the theory of systematic sampling found in Kish (40) and in Hansen, Hurwitz, and Madow (35) are useful, but the most careful and complete discussion is in Cochran (14, Ch. 8).

Lists. The discussion in Kish (40, Sec. 2.7) parallels the discussion in this chapter and provides some additional illustrations.

Use of Telephone Directories and Random Digit Dialing. Additional details may be found in papers by Glasser and Metzger (31) and by Sudman (80) in the *Journal of Marketing Research.*

Cluster Sampling

The researcher with limited financial resources will (and should) be on the lookout for sampling methods that will yield the most information for the funds available. One such possible procedure is cluster sampling. Cluster sampling is used to save survey costs and to make the data-gathering procedure more efficient. In cluster sampling, one takes advantage of the fact that the units of the population are often found in close geographic clusters. Sampling these clusters rather than individuals in the population can reduce some of the major cost components. We shall first discuss these cost components and then point out their implications for using cluster samples for various data-gathering procedures. The use of clustering is limited since, as the clusters get larger, the sampling error increases; in Section 4.3, we shall discuss why this happens, and conclude the chapter with a discussion of methods for determining optimum cluster designs.

4.1 KINDS OF CLUSTERS

Table 4.1 gives some examples of population elements and possible natural clusters of these populations. The examples given represent only some of the clusters that might be used for the populations. Deciding how to cluster may

Table 4.1
Examples of Kinds of Clusters

Population elements	Possible clusters
U.S. adult population	counties localities census tracts blocks households
College seniors	colleges
Elementary school students	schools
Manufacturing firms	counties localities plants
Airline travelers	airports planes
Hospital patients	hospitals

often be more important than deciding how large the clusters should be. The decision for using any given kind of cluster depends on the following factors:

1. The clusters must be well-defined. Every element of the population must belong to one and only one cluster.
2. The number of population elements in each cluster must be known, or there must be available at least a reasonable estimate.
3. Clusters must be sufficiently small so that some cost savings are possible—otherwise, the point of clustering is lost.
4. Clusters should be chosen so as to minimize the increase in sampling error caused by clustering.

It is not necessary that clusters be defined identically everywhere. In sampling individuals or households in urban areas, clusters will usually be blocks or groups of blocks, while in rural areas, the clusters will be geographic segments bounded by roads and natural boundaries, such as rivers and lakes. Some clusters may be single rural counties and others may be Standard Metropolitan Statistical Areas (SMSAs) consisting of several counties or, in New England, of portions of counties.

Neither is it necessary that all the clusters be the same size. In general, natural clusters will vary enormously in size, particularly as the average size of the clusters increase. Large variations in cluster sample size increase the sampling errors in surveys, but methods have been found to control this variation; these are discussed in Chapter 7.

4.2 THE COSTS OF GATHERING SURVEY DATA

The direct costs of interviewer time spent during the interview account for only a small part of the total data collection costs of face-to-face interviews. In addition to these direct costs, the hiring, training, and supervision of interviewers and the sampling costs are of major importance. Even for direct interviewer costs, a careful analysis indicates that only about one-third of the costs are for the actual interview, with about half the cost due to travel and waiting time, and the remaining costs to other clerical activities (76).

HIRING AND TRAINING COSTS

The major use of cluster sampling is the reduction of costs unrelated to the actual interview. Given a widely dispersed geographic sample, there are two procedures for reaching respondents. Interviewers may be hired and trained at one or a few central locations and sent out from there to contact respondents. The alternative is to hire and train interviewers in many locations so as to reduce the travel costs. This procedure, however, increases hiring and training costs. In the United States, which spans 3000 miles, local interviewers are used by almost all national field organizations. In smaller European countries, such as Germany, where travel costs are relatively low and it is possible to travel anywhere in a few hours, central interviewing staffs are used. In either case, it is clear that clustering can substantially reduce costs.

If interviewers are hired and trained locally, each place at which interviewers are located requires an initial trip of a field supervisor for hiring, and subsequent trips for training, supervision, and replacement of interviewers who quit. It is obviously inefficient to spend hundreds of dollars to hire an interviewer who will do only one or two interviews. If interviewers are sent from a central location to some distant interviewing point, it is obviously more efficient to do several interviews at the final destination rather than just one or two.

This discussion of costs suggests when cluster sampling is not needed. If the entire data-gathering operation is to be done by mail, using an available list, then there is no advantage to clustering. If there are to be face-to-face follow-ups of nonrespondents, however, a cluster design is more efficient even if only a small subsample are to be contacted in person. With the use of telephone interviewing, clustering may or may not be efficient, depending on the method and equipment available. If national WATS lines are used and all interviewing is from a central location, clustering is obviously unnecessary. It is quite efficient, however, if phone interviews are done by local interviewers working from their homes or from many local offices.

Clustering is primarily an efficient procedure for face-to-face interviewing. Even for this method, however, there are situations for which no clustering is required. If the total population is located in a small geographic area, such as a neighborhood in a large city or in a smaller city, and a heavy sample is selected, travel costs will be small even if simple random samples are used.

INTERVIEWER TRAVEL COSTS

If interviewers are hired locally, they usually are required to travel from their homes to sample households anywhere in their county. Since interviewers are normally paid for all travel time, including to and from their homes as well as between sample households, and are also paid a mileage allowance when using their cars, the within-county travels costs are an important fraction of direct interviewing costs. In large metropolitan areas, and particularly in the central cities, interviewers cover smaller geographic areas, but the greater density of traffic or the use of public transportation slows travel time and again makes travel costs important.

In rare cases, local interviewers may be asked to travel out of their home counties and even to spend the night at a more distant sampling point. The added costs of this travel are very high, so this is done only in emergencies, until a new local interviewer can be hired. Since counties are geographic areas about as large as can be efficiently covered by an interviewer, they usually are selected as the first-stage clusters for sampling and are called the "primary sampling units" (PSUs).

Additional clustering within the counties can substantially reduce interviewer travel costs. If more than one interview can be done at the same time on the same block, the travel costs to and from that block are spread out over several cases and reduce the cost of any one interview. Since there is a high probability that the selected respondent may not be home or available for interviewing when the interviewer calls, clustering at the block level increases the probability that at least some of the respondents will be available for interviewing. One of the most expensive and frustrating parts of the interviewing process is an interviewer's traveling 20 or 30 miles one way for an interview and finding no one home, the entire round trip wasted.

SAMPLING COSTS

Although sampling costs are not as large as interviewing costs, they are also reduced by clustering. For each place selected in a sample, one or more maps must be obtained to locate the specific area to be visited. Although no single map is very expensive, the cost of all the maps required for a national sample reaches thousands of dollars, even with clustering. Without clustering, the cost of purchasing and storing these maps is many times larger.

Even more costly is the work that must be done after places, blocks, and geographic segments are selected. Unless city directories are available, listers must go into the blocks and segments and list all the housing units in them, so that a sample may be selected. Obviously, the cost of listing is a function of the number of clusters, not of the sample size. The cost of listing a block remains the same whether 1, 5, 10, or 20 units from that block are selected later for interviewing.

4.3 INCREASES IN SAMPLE VARIABILITY DUE TO CLUSTERING

If cost were the only concern, the logical conclusion would be that all interviewing should be done in one or a few giant clusters. The idea that there is one or a few counties that are representative of the United States has often been expressed, particularly during national elections. Some reporter is always discovering a county that has voted like the total country in the last *n* presidential elections. There is no reason to believe, however, that this county will continue to be representative. Earlier in the twentieth century, there was the saying "As goes Maine, so goes the nation." After the 1936 election, this saying was revised to "As goes Maine, so goes Vermont," since these were the only two states to vote for Landon instead of Franklin Roosevelt.

Even if a county appears typical in some characteristics, such as age or income distribution, it cannot be typical for all variables. Even more important, no single county or even a few counties can capture the variability that is part of the essence of the United States.

Intuitively, we can sense that a sample of 2000 households selected from one or two counties is not the same as a sample of 2000 households selected from 200 counties, but how do they differ? The important thing to consider is not the total sample size but the number of independent observations provided by the sample. Since the notion of independence is difficult, the following example may help to clarify the basic idea.

Example 4.1 A Racially Segregated Neighborhood

Suppose one wishes to estimate the proportion of black and white residents in a neighborhood. The cluster sample design calls for a selection of 50 blocks with 20 households per block, or a total sample of 1000 households. Once in the neighborhood, it is learned that all blocks are completely segregated by race. That is, the blocks are either all black or all white. In this case, how many independent observations does the sample yield?

If the interviewer knocks on the door of the first house of the block and the respondent is white, all other households on that block will also be white; if the respondent is black, all other households will be black. It is clear that there is really only *one* independent observation per block, so only 50 independent observations from the 50 blocks.

Suppose, on the other hand, that the same sample design and blocks are used, but now the purpose of the study is to estimate birth patterns for the year to discover if births are more likely to occur in some months than in others. As far as we know, there is no relation between race and birth patterns, nor do we expect any relation between the birthdates and neighbors. In this case, the sample consists of 1000 independent observations.

The example illustrates the important fact that the amount of clustering that should be done depends on the homogeneity of the cluster: The more homogeneous the cluster, the less information is obtained as the cluster sample becomes larger. Note that the homogeneity of the cluster depends on the variable being considered—the clusters in the sample are completely homogeneous racially but heterogeneous with respect to birthdate.

MEASURING OR ESTIMATING HOMOGENEITY

Once the study is completed, it is possible to determine the homogeneity within clusters; but, by that time, of course, it is too late for optimum sample planning. For planning purposes, one uses estimates of homogeneity based on previous studies for similar topics. The standard measure of homogeneity is ρ (*rho*), which behaves like a correlation coefficient. It is the correlation between all the possible pairs of elements within the cluster. If the cluster is completely homogeneous, $\rho = 1$; when the homogeneity equals that of random sorting of elements into clusters, $\rho = 0$. If there is extreme heterogeneity within the cluster, $\rho = -1/(\bar{N}-1)$ where $\bar{N}$ is the average cluster size. If $\bar{N} = 2$, ρ can be -1, so it has the same range as other correlation coefficients. Negative values of ρ are seldom observed, however, except in observing the sex of the respondent within the household. That is, if two interviews are taken from the same household, there is a high likelihood that the sexes of the respondents will be different. This obvious result indicates that, if one were interested in comparing differences in attitudes or behavior between men and women, clusters of two respondents per household, one male and one female, would be an optimum design.

In the typical case, ρ is positive and increases as the units within the cluster become more homogeneous. As one might expect, the greatest homogeneity is usually found for economic variables, since these determine the locations households can afford. Even more homogeneous are the housing characteristics of the neighborhood, such as the value of owner-occupied homes and the monthly rent in rental units. Since most social variables are influenced to some extent by economic variables and since housing choices reflect other similarities among households, some degree of homogeneity is found for almost all variables—although, in many cases, ρ is very small and approaches zero.

Appendix B gives some values of ρ for selected demographic and health variables and cluster sizes. These results may be used for planning future cluster samples. Several aspects typical to cluster sampling are evident in this appendix. In Table B, note that clusters of farms yield especially high values of ρ due to local similarities in soil and climate conditions by geography. In the household samples in Tables A and D, as discussed earlier, homogeneity is highest for rent paid and household income. A relatively large value of ρ is observed for average size of household since this variable determines the amount of housing space required and clusters are similar in the size of housing offered.

In Table D, we see that the highest homogeneity is that of the racial distribution within the block. While ρ is not 1.0, as in Example 3.1, it ranges up to .6, which is three times larger than any other values of ρ. In general, values of ρ for health statistics are small, averaging around .05 or lower, even though information is obtained for all members in a household.

Note that, as the cluster sizes increase, the homogeneity becomes smaller; that is, larger geographic areas become increasingly more heterogeneous. This might suggest that larger clusters are better than smaller ones for reducing sampling error, but cost factors must also be considered. As clusters increase in geographic area, travel costs increase if only a few elements are chosen from the larger cluster. If more elements are chosen in the larger cluster, then sampling errors increase. This relation is discussed in greater detail in Sections 4.4 and 4.5.

The smallest values of ρ are observed for medical variables in the clusters of employee groups in Table C of Appendix B. It will generally be true that nonhousehold clusters will not be homogeneous with respect to variables unrelated to the group's purpose. That is, work groups will be homogeneous on working conditions but not other variables; school groups will be homogeneous on education variables, but not others. In addition, the values of ρ are small for medical variables even when geographic clusters are used, as seen in Table D.

If there are no previous data available for the variables under study and if one is afraid to make estimates of ρ based on other variables, such as those in Appendix B, it is always possible to conduct a pilot study to estimate ρ. This is just one example of the use of a pilot study for improving the design of a larger study. Other examples are given in Chapter 6, which deals with stratified sampling. Pilot studies do add to cost, and delay the start of the main study, so they should not be used to improve sampling efficiency unless other sources are unavailable or unsatisfactory.

4.4 THE EFFECTS OF CLUSTER SIZE ON SAMPLING ERROR

As was discussed in the previous section, clustering reduces the number of independent observations and increases the sampling error. The increase in sampling error is a function of the size of the cluster and of ρ, the homogeneity within the cluster. Hansen, Hurwitz, and Madow first proved the extremely useful formula that relates cluster samples with simple random samples when the sample size is the same.

They showed that, if $\bar{n}$ is the average size of the sample from each cluster, then the ratio of the sampling error for a cluster sample of size n to that of a simple random sample of the same size is:

$$\frac{s^2_{\text{cluster}}}{s^2_{\text{simple random}}} = 1 + \rho(\bar{n}-1). \tag{4.1}$$

Although the formula is only approximately true, the approximation is sufficiently accurate for practical use if the clusters are about the same size. The formula does assume that the number of clusters in the sample is small, relative to the total number of clusters in the population. If this is not the case, a finite correction factor is applied, thus reducing the cluster sampling error. The finite correction factor is discussed more generally in Chapter 8.

Formula (4.1) may be used in several ways. If ρ is known or can be estimated, the cluster sampling error can be estimated, since the simple random sampling error is either known or can be computed easily. On the other hand, if the cluster sampling error is known, ρ can be estimated, by rearranging the terms in Formula (4.1):

$$\rho = \frac{s^2_{\text{cluster}} - s^2_{\text{simple random}}}{(\bar{n} - 1)\, s^2_{\text{simple random}}}. \tag{4.2}$$

The ratio of the variance of *any* sample design to the variance of a simple random sample of the same size is often called the "design effect," sometimes abbreviated as "deff." The ratio of the variances of samples with the same design effects will depend only on the ratios of their sample sizes. Although the idea of design effect is useful in the planning of complex surveys, it does depend on cluster sample size, so a change in the clustering procedure changes the design effect.

A researcher would usually compute the cluster sampling error directly at the end of a study, using the methods discussed in Chapter 8 for estimating sampling errors from complex samples. Next he would compute ρ from Formula (4.2) and use this estimate of homogeneity for planning future studies. An example using the data in Appendix B will illustrate the process.

Example 4.2 Planning a Medical Expenditure Study

In the Anderson (3) study cited in Table C of Appendix B, the total sample was 360 cases consisting of 18 companies, and $\bar{n} = 20$ employees per company. For net medical charges for physician services, the estimated sampling error was \$5.10. The sampling error for a simple random sample of 360 cases was estimated as \$3.50. It may be seen that the cluster sample had a sampling error 1.5 that of the simple random sample of the same size. Using Formula (4.2),

$$\rho = \frac{5.1 - 3.5}{19\,(3.5)} = \frac{1.6}{66.5} = .024.$$

Now, in the planning of a new study using employee groups, this estimate of ρ could be used. There is a different value of ρ for each variable, but assume that this is one of the key variables in the new study or that it is used because it indicates the homogeneity most likely to be found.

Suppose that cluster sizes of 3, 5, 8, 10, 15, 20, and 25 are being considered. Using Formula (4.1), it is easy to compute the ratio of the cluster sampling error to that of simple random sampling (see Table 4.2).

Looking only at the increase in variance, one would conclude that the smallest cluster sizes are most desirable, but it is also necessary to remember costs. Suppose that the cost per case of a simple random sample is \$50, the cost for clusters of 5 is \$30 per case, and the cost for clusters of 25 is \$15 per case. Which sampling plan gives the most information given a budget of, say, \$6000? The size of the simple random sample is 120 units; the size for the cluster sample with clusters of 5 is 200 units in 40 clusters; the size for the cluster sample with clusters of 25 is 400 units in 16 clusters. Now, assuming the same value of ρ so

Table 4.2
The Ratio of Cluster to Simple Random Sampling Error as a Function of Cluster Size

Cluster size	$\frac{s^2 \text{ cluster}}{s^2 \text{ simple random}}$
3	1.048
5	1.096
8	1.168
10	1.216
15	1.336
20	1.456
25	1.576

Table 4.3
Sampling Variances and Equivalent Independent Observations
for Alternative Cluster Sizes

Sample design	Actual n	$\frac{s^2_{\bar{x}} \text{ cluster}}{s^2_{\bar{x}} \text{ simple random}}$	$s^2_{\bar{x}}$	Equivalent independent observations
Simple random	120	1.000	$s^2/120$	120.0
Clusters of 5	200	1.096	$1.096s^2/200$	182.5
Clusters of 25	400	1.576	$1.576s^2/400$	253.8

that the ratios in Example 4.2 hold, one can compute the sampling variances for the three designs, assuming a universe variance of s^2.

It may be seen in Table 4.3 that the sample of 200 with clusters of 5 is really equivalent to a simple random sample of size 200/1.096 or 182.5, and the sample of 400 with clusters of 20 is equivalent to a simple random sample of size 400/1.576 or 253.8.

In this case, it is evident that, considering only these three alternatives, the cluster sample with clusters of 25 has the smallest sampling error and therefore gives the most information for the funds available. More generally, the optimum cluster size depends on the data-collection cost function and ρ, the homogeneity within clusters. Up to the optimum cluster size, costs decrease more rapidly than does sampling variance; while beyond the optimum cluster size, the increase in variance is greater than the decrease in cost. In Section 4.5, the general solution for finding the optimum cluster size is given, based on some simple cost functions.

4.5 OPTIMUM CLUSTER SIZES

The major decisions made in planning a study are the number of primary sampling units or areas (PSUs) and the average cluster size per PSU. A simple cost function for making this decision is:

$$C_t = ac_1 + nc_2 \tag{4.3}$$

where

C_t = total cost of the study
a = number of PSUs
c_1 = average PSU costs allocated to a given study
n = total sample size for a given study
c_2 = average unit costs per interview

The c_1 costs are those of hiring, training, and supervising the interviewing staff at each PSU, as well as the sampling costs. The c_2 costs are the direct interviewer costs. Note that this cost function deals only with interviewing costs and does not consider other survey costs, such as data processing and analysis. These other costs are irrelevant when computing an optimum cluster size but must be taken into account when estimating the total budget.

While this cost function is highly simplified, it is sufficient for determining an optimum cluster size and thus the optimum number of PSUs. Hansen, Hurwitz, and Madow proved that, given this cost function,

$$\bar{n}_{\text{opt.}} = \left[\frac{c_1}{c_2}\left(\frac{1-\rho}{\rho}\right)\right]^{1/2} \tag{4.4}$$

Given Formulas (4.3) and (4.4), the number of PSUs is determined if either the total sample size n or the total cost C_t is specified. If n, the total sample size, is specified, then a, the number of PSUs, is $n/\bar{n}_{\text{opt.}}$. If C_t is specified, then Equation (4.3) may be rewritten as

$$C_t = ac_1 + a\bar{n}c_2 \tag{4.5}$$

and solving for a,

$$a = \frac{C_t}{c_1 + \bar{n}_{\text{opt.}}\, c_2}$$

As in estimating ρ, the values of c_1 and c_2 are best determined from previous experience. Although exact values of c_1 and c_2 are dependent on current wage rates, it has been observed in several national survey organizations that the ratio of c_1 to c_2 ranges from about 25 to 50.

Example 4.3 Estimating the Optimum Number of PSUs for a Multipurpose National Sample

A survey organization wishes to set up a new sample that will be used for a variety of social science studies. While values of ρ cannot be well-specified, it is estimated that most will be in the range of .05 to .10; the range c_1/c_2 is estimated as between 25 and 50. Using Formula (4.4), the optimum $\bar{n}$s for these ranges are:

$c_1/c_2 = 25$	$\rho = .05$	$\bar{n}_{\text{opt.}} = 22$
	$\rho = .10$	$\bar{n}_{\text{opt.}} = 15$
$c_1/c_2 = 50$	$\rho = .05$	$\bar{n}_{\text{opt.}} = 31$
	$\rho = .10$	$\bar{n}_{\text{opt.}} = 21$

The decision is made to let $\bar{n}_{\text{opt.}} = 20$. Although this will not always be the optimum, the efficiency of most samples declines very slowly in the neighborhood of the optimum values so this $\bar{n}$ will be near optimum for most purposes.

If a total sample of 2000 cases is desired and if c_1 is estimated at \$600 and c_2 at \$20, then there will be 100 PSUs and the total cost will be 600(100) + 2000 (20), or \$100,000. In fact, many national survey organizations do have about 100 PSUs in their national samples, with samples of 15 to 30 cases per PSU.

Suppose, on the other hand, that a total budget of \$25,000 is allowed for a study. Then, from Equation (4.5),

$$a = \frac{25{,}000}{600 + 20 \times 20} = 25.$$

Thus, here the total sample size n would be 500 cases in 25 PSUs.

Example 4.4 The Optimum Number of PSUs for a State or Regional Sample

The major difference between a national and a state or regional sample is that the c_1 costs are lower for regional samples because less travel time and expense are required. Often the supervisor who hires and trains interviewers can go and return in the same day. The ratio of c_1 to c_2 is between about 15 and 20 for many state and regional samples. Using values of ρ between .05 and .10 as before, the optimum $\bar{n}$ is between 12 and 19. Selecting an average $\bar{n}$ of about 15 units would be near optimal for many purposes.

If only a single county or several contiguous counties are of interest for some local policy issue, then there is no clustering of counties, of course, although there may be some clustering within counties. If, on the other hand, one or a few counties are chosen, either judgmentally or by some probability procedure, to represent the total U.S. population, then, even if ρ is very small, the sampling variance will be very large and the design may be far from optimum.

For some very small studies, it may be possible to justify extremely heavy clustering if one considers only out-of-pocket expenses related to PSU costs c_1 and interviewing costs c_2. Thus, a graduate student or individual researcher might do all the interviewing and other activities related to a project, and consider this time as completely or virtually cost-free. In this case, only travel expenses would be of real importance, and the ratio c_1/c_2 would become very large. Even when heavy clustering is justified this way for very small studies, it should be recognized that not only will the sampling variances be large but any estimates of sampling variances will themselves be subject to very large variances. At a minimum, it is necessary that at least two PSUs be selected to make it possible to make any estimate of sampling variance (see Chapter 8).

Example 4.5 The Optimum Cluster Size for a Continuing Study

If interviewers can be hired and trained for a repetitive study, the c_1 costs can be spread over the expected length of the interviewer's employment. Assuming monthly interviews, as in the Census Bureau's Current Population Survey, and assuming the average interviewer stays about a year, the ratio of c_1 to c_2 drops to about four. In this case, assuming ρ in the range of .05 to .10, the optimum cluster size is six to nine per PSU. A standard procedure used by the Census Bureau is to make the cluster size in a PSU the amount of work that can be accomplished by a single interviewer. It is clear that this is a near-optimum procedure for repetitive studies; but it should also be noted that heavier clustering is optimum for nonrepetitive studies, in which interviewer training is required for each new study and interviewer turnover is higher because of the uncertain workload.

Example 4.6 The Optimum Cluster Size for a Study of Graduate Students

A large mail sample of graduate engineering students was conducted during the 1960s to determine the amount of financial support the students needed and were getting. There might seem to be no reason for clustering in this case, but it was necessary to enlist the support of a faculty member at each selected university to obtain the required lists of students and to follow up on nonrespondents. An honorarium of $200 was paid to the representative at each school. Since the mailing costs were about 50¢ a student, $c_1/c_2 = 400$. Values of ρ between .10 and .20 were expected within a given field. Thus, the optimum cluster size per school per field was expected to be in the range of 40 to 60, and an average $\bar{n}$ of 50 was chosen as near optimum.

4.6 OPTIMUM CLUSTER SIZES WITHIN PSUs

The same factors discussed in Section 4.5 must be considered in determining optimum cluster size within PSUs. Clustering at the block or segment level reduces interviewer travel time and costs. This is especially true in areas like large cities, where it is difficult to find respondents at home. If an interviewer has only one assignment on a block and that respondent is not home, the entire trip is wasted; if, on the other hand, there are several assignments on the block and one respondent is not home, the interviewer is more likely to find someone else in the sample cluster. Since blocks tend to be economically homogeneous, however, clustering at the block level may also increase sampling errors.

To determine an optimum cluster size at the block or segment level, a more complex cost function is required. This cost function will not be discussed here since no simple solution for using it is possible, but it is discussed in Hansen, Hurwitz, and Madow (35, Ch. 6.18) where a method is also given for finding an approximate solution. For many social science variables, clusters of 3 to 8 households per block or segment are near optimum, except for race variables, as noted in Appendix B. The larger size clusters may be used if variables are unrelated or only slightly related to economic variables. Much larger clusters may sometimes be optimum if not all households qualify and screening is required to identify those eligible for interviewing. Thus, if only 1 household in 10 is eligible, clusters of 50 may be near optimum. Procedures for locating rare households are discussed in greater detail in Chapter 9.

Once a cluster size per block or segment is determined, the number of blocks is also determined as the ratio of PSU sample size to cluster size per block.

Example 4.7 Optimum Number of Blocks or Segments per PSU

Earlier, we decided to have about 20 cases in each PSU. If we now decide to have clusters of about five per block or segment, there will be four blocks or segments; if we take about four households per block, there will be five blocks per PSU.

4.7 CLUSTERING WITHIN THE BLOCK OR SEGMENT

Within the block or segment, two methods of sampling for households are used. The more common procedure is to take a systematic sample of every *n*th household. Since adjacent households are most alike, the systematic sample generally yields a ρ value slightly smaller than that obtained by a procedure using compact clusters in which a group of adjacent households all are interviewed. The systematic sample also reduces the possible contamination that may occur if neighbors talk to each other about the survey and influence each other's responses.

For use in a continuing survey or a multipurpose sample, the compact cluster has the advantage that *new* housing units are automatically included if they fall within the cluster boundaries, and that the initial listing may be done more thoroughly. In using a systematic sample within the block or segment, it is possible for the interviewer to note new construction, but there is a greater likelihood that some units may be overlooked.

CLUSTERING WITHIN THE HOUSEHOLD

When the household is the unit of investigation, there is, of course, no clustering within it. If, however, the population is of all persons or all adults, one must decide on the number of persons to include in each household. For some studies, such as the National Health Survey, information about all household members is obtained from one household informant. Although this increases the variance over a nonclustered sample of individuals, as may be seen in Appendix B, this is a very efficient procedure because the cost of obtaining information about all household members is only slightly higher than that of obtaining information about only the respondent.

If information must be obtained from each individual household member, costs increase considerably, along with the possibility of contamination of responses as individual members discuss the survey with each other. Except when one wishes to compare the responses of husband, wife, and other family members, the general procedure is to choose only one or two respondents per household.

If this procedure is followed exactly it leads to a biased sample since members of smaller households have a higher probability of selection than do members of large households. In particular, in one-member households, the respondent is selected with certainty. This leads to an oversampling of widows and older women and the distortion of the demographic characteristics, as well as other variables, of the sample. One solution is to take a sample of $1/n$th of the members of each household. Generally, when sampling adults, the sampling fraction is 1/2 because typically there are two adults per household. This means that half of the one-member households are not interviewed. Among those households with three adults, there would be one interview half the time and two interviews the other half. Households with four adults will always have two interviews. Since less than 1% of households have more than four adults, it is a common procedure to sample no more than two adults in any household. Usually, the bias caused by missing a few adults in households with five or more adults is small enough to be ignored.

Some researchers never take more than one interview per household, but this biases the sample against that third of the adult population who live in households of three or four adults. Since these households are usually those with children over age 18 who have not yet set up their own households, this procedure shifts the sample age distribution toward older respondents, even when half of the one-member households are omitted. One other method for correcting the bias that results from one interview per household is to weight the results by number of eligible household members.

4.8 SUMMARY

To obtain an optimum cluster design, the cost savings of cluster samples must be balanced with the increase in sampling variability. The increase in variability is due to homogeneity within the cluster and can be estimated from previous studies of similar variables. Similarly, costs of interviewer hiring and training, travel time, and actual time spent in interviews should be estimated from past experience. There are no reasons for using clustering if data collection is done by mail or phone from a central location. Given estimates of ρ—the homogeneity within the cluster—and a cost function, optimum or near-optimum cluster sample sizes can be determined.

It is frequently observed that survey organizations use clusters of 15–30 per county or PSU for national samples. Smaller cluster sizes are optimum for local and regional studies and for panel studies. Within the PSU, clusters of three to eight households per block are near optimum for many social science variables.

4.9 ADDITIONAL READING

An excellent detailed discussion of clustering and cost functions is given in Chapter 6 of Hansen, Hurwitz, and Madow (35). Some empirical data on survey costs are given in Sudman (79, Ch. 2). These data must be adjusted for increases in the price level since they were collected, but relative allocations should still hold.

Readers who are interested in the methodological details of how the intra-cluster correlations in Appendix B were estimated should consult the report by Waksberg, Hanson, and Jacobs for the U.S. National Center for Health Statistics (96). This report also gives some additional correlations for population sub-groups that are not included in Appendix B.

How Big Should the Sample Be?

5.1 INTRODUCTION

One of the first questions that confront the designer of a new study is "How big should the sample be?" Although it appears to be a simple, straightforward question, it is one of the most difficult to answer precisely.

There are several ways of approaching this problem. The easiest is an empirical approach, discovering what sample sizes have been used by others with similar problems. It is useful, especially for an inexperienced researcher, to check his judgment about appropriate sample sizes against those of other social scientists. Section 5.2 of this chapter presents actual sample sizes used in a large number of studies.

The remainder of the chapter is devoted to a more formal approach, which emphasizes the need to balance the value of increased information with the costs of gathering the data. The problem is considered from the viewpoints of both the designer of a study and the funding organization.

The designer of a study must recognize that, if the study is to be funded by someone else, the organization funding the study will ultimately make the decision about sample size. This chapter offers suggestions to aid the funding agency in this decision process, and proposes strategies that may be helpful to the writer of a research proposal when suggesting a sample size.

Although the precise benefits of most basic social science research cannot be well-specified, an optimum sample size can be determined for business and policy applications if gains and losses of alternative decisions are known. A simple, linear, two-decision problem is discussed to illustrate how the value of information is determined.

One important question that is frequently overlooked entirely is whether or not new sample information should be gathered *at all.* The decision depends on both the cost and the value of the new information. If the fixed costs of the study exceed the value, no new sampling is justified. It will be shown that the value of information depends on what is already known and is, thus, best described by using Bayesian statistics rather than the classical statistics with which most readers are familiar. Although every effort is made to keep this discussion simple, the reader who has been exposed only to classical statistics may find Sections 5.6–5.11 difficult, and may skip them on first reading or concentrate only on getting the major ideas.

5.2 CURRENT SAMPLE SIZES USED

In this section are summarized the sample sizes of several hundred studies used in a literature review of response effects in surveys (81). Although, in some cases, the sample sizes used appeared inappropriate for the purposes, the modal sample sizes reflect current practices and are a useful guide for planning. Table 5.1 is not intended to replace the formal procedures for sample size determination described later in this chapter, but to give the inexperienced researcher some way of checking his judgments against those of others in the same field. Since Table 5.1 is split into broad subject matter areas, a researcher in a specialized field can replicate these results by noting the sample sizes reported in the specialized journals for the field. In addition to the modes, the seventy-fifth

Table 5.1
Most Common Sample Sizes Used for National and Regional Studies, by Subject Matter

	National			Regional		
Subject matter	Mode	Q_3	Q_1	Mode	Q_3	Q_1
Financial	1000+	–	–	100	400	50
Medical	1000+	1000+	500	1000+	1000+	250
Other behavior	1000+	–	–	700	1000	300
Attitudes	1000+	1000+	500	700	1000	400
Laboratory experiments	–	–	–	100	200	50

Table 5.2
Typical Sample Sizes for Studies of Human and Institutional Populations

	People or households		Institutions	
Number of subgroup analyses	National	Regional or special	National	Regional or special
None or few	1000–1500	200–500	200–500	50–200
Average	1500–2500	500–1000	500–1000	200–500
Many	2500+	1000+	1000+	500+

and twenty-fifth percentiles are also given to indicate the range of sample sizes.

It may be seen that national studies, regardless of subject matter, typically have samples of 1000 or more. Regional studies vary considerably, depending on the topic, but, as expected, usually have smaller samples. The topic of the study is not really the basic factor that determines sample size. As indicated earlier, sample size depends on how many population subgroups one wishes to study. Table 5.2 gives some typical sample sizes when few or no subgroups are to be analyzed and when many subgroups are of interest. There are also some sample sizes for populations that consist of institutions or firms rather than of people or households. These sample sizes are given with the assumption that the optimum stratified sampling procedures discussed in Chapter 6 are used. The sample sizes for institutions are generally smaller because of these optimum procedures and because the sample is frequently a large fraction of the total population. Like Table 5.1, Table 5.2 should be used as an initial aid rather than as a substitute for formal judgment.

5.3 SAMPLE SIZE DETERMINATION IN SPECIAL CASES

In this section are discussed some special cases that are frequently encountered but that do not fit the general suggestions, given earlier, for determining sample size.

PILOT OR PRE-TESTS

Pilot tests are an important step in developing survey instruments. It is extremely difficult even for experienced social scientists to write a questionnaire with no confusing or ambiguous questions. A pilot test of 20–50 cases is usually sufficient to discover the major flaws in a questionnaire before they damage the main study. The larger sample is required if there is concern that some subgroups in the population, such as those who are retired or who have less education, will have difficulties.

COSTS FIXED

It is sometimes the case that an amount of money M is allocated for a project before any decision about sample size is made. The study director then has two decisions–how to allocate the funds between data collection and data analysis, and whether or not to do the study at all. It is evident that spending all the money for data collection would leave none for analysis and spending it all on analysis is impossible since there would be nothing to analyze. A rough rule of thumb observed in many studies is that about half the money is spent for data collection and the other half for data analysis. This rule appears to be based on the judgment that data collection and analysis are of about equal importance.

Suppose that the decision is made to allocate $M/2$ dollars for data collection. This determines the sample size, once the data-collection procedure is specified. In the profit sector, the decision maker should consider whether or not the reduction in uncertainty about the decision results is worth the cost. When conducting basic research, the researcher must decide if the funds are sufficient to enable him to achieve at least some, if not all, of his research goals. If not, the funds should be rejected. An especially nasty situation arises when both the costs and the accuracy requirements are specified; frequently, it is not possible to meet the accuracy requirements with the given costs. Obviously, here the problem is not in finding a sample size but in negotiating with the funding group so that the requirements are modified. That is, either the cost constraint or the accuracy constraint must be abandoned, or they may be jointly modified. If this is not possible, the study should not be conducted.

NONMONETARY COSTS

Suppose no out-of-pocket expenses are required for data collection, but the work will be done by the researcher, himself, or by others who receive no pay. If the time of the data gatherers is considered an economic resource, sample size should be determined in the manner just discussed. The researcher and his staff should consider the value of the data and the alternative uses of their time. It is obvious that, if the sample becomes too large, it will require more time than can be spared for a single project. If a given amount of time T is available for the research, the rule of thumb given for allocating fixed funds would appear to apply: About half the time should be spent in data collection and half in data analysis. As with funded studies, the researcher must decide whether or not to do the study at all, given the amount of time available and the value of the information.

A common example of this situation is doctoral dissertation research in which the data are gathered by the doctoral candidate. There is frequently a conflict between the candidate and his committee over sample size, since, for the committee, the data are free, but, for the candidate, they involve the use of a scarce resource–time. Usually, the compromise involves a joint decision on the amount of time to be spent in data collection.

5.4 COMPARISON OF CLASSICAL AND BAYESIAN PROCEDURES FOR SAMPLE SIZE SELECTION

We now turn to a more formal analysis of methods for deciding sample size. It is likely that some readers who have been exposed to classical statistical procedures may remember that there is an exact formula for computing a required sample size if the width of the required confidence interval and the probability level are specified. But it is obvious that this formula merely shifts the problem from that of specifying *sample size* to that of specifying the *width of the confidence interval,* the substitution of one problem for another.

Perhaps the most sophisticated statement of the traditional approach to sample size determination is by Cochran (14). He outlines the following steps:

1. There must be some statement concerning what is expected of the sample. This statement may be in terms of desired limits of error . . . or in terms of some decision that is to be made or action that is to be taken when the sample results are known. The responsibility for framing the statement rests primarily with the persons who wish to use the results of the survey, though they frequently need guidance in putting their wishes into numerical terms.
2. Some equation which connects n with the desired precision of the sample must be found. The equation will vary with the content of the statement of precision and with the kind of sampling that is contemplated. One of the advantages of probability sampling is that it enables this equation to be constructed.
3. This equation will contain, as parameters, certain unknown properties of the population. These must be estimated in order to give specific results.
4. It often happens that data are to be published for certain major subdivisions of the population and that desired limits of error are set up for each subdivision, and the total n is found by addition.
5. More than one item or characteristic is usually measured in a sample survey: sometimes the number of items is large. If a desired degree of precision is prescribed for each item, the calculations lead to a series of conflicting values of n, one for each item. Some method must be found for reconciling these values.
6. Finally, the chosen value of n must be appraised to see whether it is consistent with the resources available to take the sample. This demands an estimation of the cost, labor, time, and materials required to obtain the

> proposed size of sample. It sometimes becomes apparent that n will have to be drastically reduced. A hard decision must then be faced–whether to proceed with a much smaller sample size, thus reducing precision, or to abandon efforts until more resources can be found [pp. 72–73].[1]

But, as pointed out in Cochran's Step 6, it is often the case that the project director's desires for accuracy conflict with the resources available, and some method must be found for resolving this conflict. The Bayesian approach, described later, is thus not in conflict with the classical approach, but is an attempt to apply more formal methods to a very complex problem. By decomposing the sample size decision into more basic elements, one pinpoints the critical ones for which estimates must be improved. Section 5.5 discusses the most critical element in determining sample size—the value of information.

5.5 THE VALUE OF INFORMATION

The purpose of sampling is to obtain information either for basic research or for decision making by either a profit-making or nonprofit organization. Information is essential if organizations or societies are to survive in a complex civilization—it is information, not love, that makes the world go round.

Enthusiasm for information, however, is not enough. We all are sometimes aware that there can be too much of a good thing. Both the decision maker who declares, "Don't bother me with the facts, I've made up my mind," and the little girl who is unhappy with a book because "It tells me more about elephants than I wanted to know" are saying that, for them, information on certain topics is of limited value.

It seems most useful to consider information as just another economic good comparable in many ways to chocolate ice cream. Economic goods have decreasing marginal utility. After having eaten a gallon of chocolate ice cream in a restaurant, very few of us would order another gallon. Similarly, there comes the time when additional information is not worth the cost of obtaining it. Another aspect of economic goods is that they have different values for different users. Information about consumer attitudes toward beer may be of enormous importance to beer marketers, of some interest to social psychologists, and of no interest at all to neighborhood florists. Even among *users* of a good, the value will vary with the circumstances. Chocolate ice cream is in less demand at hockey games than at summer picnics. The value of information to its user depends on how likely the information is to influence an action or decision. The ideas become more clear when we view a situation in which information is used

[1] William G. Cochran, *Sampling techniques,* pp. 72–73. © 1963 by John Wiley & Sons, Inc. Reprinted by permission of John Wiley & Sons, Inc.

to aid a decision maker to decide between two choices that have economic consequences. We discuss this case first, before turning to the case of pure research.

5.6 VALUE OF INFORMATION FOR DECISIONS FOR MAXIMIZING GAINS

There are three factors that determine the value of new sample information for a decision where gains and losses can be specified:

- The current degree of uncertainty of the decision maker
- The gains and losses connected with the alternative decisions
- The effect of uncertainty on the decision

THE CURRENT DEGREE OF UNCERTAINTY

Consider the sentence "Information is essential." The value of the information it contains for the reader is probably zero, for he is already aware of the information contained in that sentence. The information that the sun rises in the east is of no value to adults, but the exact location of the sun at a given point in time is not known to most of us, and could be of value. If we know, or think we know, some fact with certainty, and no new information could make us change our mind, then there is no value to additional information. An eccentric Illinois millionaire had a standing offer of $25,000 to anyone who could provide him with information that would convince him that the earth was not flat but round. No one ever collected the reward, although several attempts were made. He went to his grave still firmly convinced that the earth was flat. For him, new information on this topic had no value.

A more usual case would be the city mayor who is so convinced that a new health program just developed by the city will be beneficial that any information obtained from research would be valueless to him. If the research results are negative, he will disregard them; if they confirm his view, he will merely do what he already had intended doing.

As our examples show, the value of information is a very personal concept, which depends on individual knowledge and beliefs. The same information may have great value for some people or organizations, and no value for others. It depends on what the person already knows. Since no two people have had exactly the same background and experiences, the same information will usually not have the same value for them. Thus, whenever we talk about the value of information, we must specify for whom.

The basic difference between classical and Bayesian statistical procedures is that the Bayesian procedures incorporate prior judgments and uncertainty into the decision process. Decisions in the profit sector and on public policy issues are made by executives who are in their important positions because of their experience and presumed good judgment. For basic research, funding decisions are made by peer review, the reviewers also having been chosen for their experience and good judgment. The formal Bayesian methods do not guarantee, however, the accuracy of prior judgments. They do enable the decision maker to use his judgment in a logical way, while with informal methods, there is at least the possibility that fuzzy thinking will distort the process.

Uncertainty is generally expressed as a variance or standard deviation around the most likely value. In classical statistics, one uses $\sigma_{\bar{x}} = \sigma/\sqrt{n}$ to indicate uncertainty about the mean after observing a simple random sample of size n. Since the decision maker's uncertainty is not based solely on sample evidence but on past experience and judgment as well, V will be used instead of $\sigma_{\bar{x}}$. Note, however, that neither V nor $\sigma_{\bar{x}}$ are the same as σ, the population standard deviation. If there is available any previous experience or sample information, V will be smaller than σ.

Example 5.1 Political Campaigning: Degree of Uncertainty

A candidate for mayor has spoken with various civic groups and with his own party workers and thinks he will win the coming election. He judges that he will get about 52% of the votes and that the chances are two out of three that he will get between 50% and 54% of the vote.

The universe value σ^2 for the binomial distribution is $p(1-p)$. Using the candidate's estimate of $p=.52$, $\sigma^2 = (.52)(.48) = .2496$ and $\sigma = \sqrt{.2496} = .5$.

If his prior distribution is normally distributed, $V = .02$, since two-thirds of a normal distribution are between plus and minus one standard deviation from the mean.

Suppose, instead of basing his judgment on his talks, he had commissioned a survey of 624 voters and discovered that 52% favored his election. From this sample data:

$$\sigma_{\bar{x}}^2 = \frac{p(1-p)}{n} = \frac{.2496}{624} = .0004,$$

and

$$\sigma_{\bar{x}} = .02.$$

Noting that $V = \sigma_{\bar{x}}$, one would conclude that the candidate's prior judgment based on informal procedures was the equivalent of the more formal procedure of taking a simple random sample of 624 voters.

Given his prior judgment, the candidate might make a series of economic decisions, such as quitting his job to campaign more vigorously, borrowing money for campaign expenses, or relaxing his campaign efforts. He might decide also that his prior judgment was insufficient aid for making these decisions and that he needed a formal poll of voters to reduce his uncertainty. If he did spend money for a poll, he would not ignore his informal sources of information but combine them with the poll results in making his final decisions.

THE GAINS OR LOSSES CONNECTED WITH EACH POSSIBLE DECISION

The magnitude of the possible gains and losses of a decision also determine the value of new information. When expenditures will run into the millions, information clearly is more important than for decisions involving a few thousand dollars. For many small decisions, no new sample information is justified because the cost of obtaining it would be greater than the cost of the decision's result.

It also follows that, other things being equal, the more persons affected by a decision, the greater the value of information. Thus, national decisions generally require more information than do regional decisions, and, similarly, state decisions require more information than do neighborhood decisions.

One of Parkinson's tongue-in-cheek laws is that the deliberation over a decision is inversely related to its size, since everyone can understand small amounts but no one comprehends huge expenditures. In a rational world, however, it would make no sense to do an expensive study of playground usage in order to determine if a new swing is needed. The cost of the study would be greater than the cost of the swing. On the other hand, for major changes in welfare services, such as health insurance for everyone or increased federal support of public education, the need for additional information is evident. For projects of intermediate importance, careful formal analysis is required to determine the worth of new information.

THE EFFECT OF UNCERTAINTY ON THE DECISION

For a simple linear two-action decision problem there will be a break-even point. If the true state of the world is above this break-even point, one course of action will be taken, if below, the other course. The nearer the decision maker's prior expectation is to the break-even point, the greater the value of information.

The decision maker's current degree of uncertainty (discussed on pp. 91–92) must be considered, relative to the break-even point. If his prior expectation is far higher or lower than the break-even point, as in Figure 5.1(a), there will be little or no value to additional information, even if there is a high degree of uncertainty. On the other hand, if the prior expectation is near the break-even point, information may be very valuable even with little uncertainty, as in Figure 5.1(b).

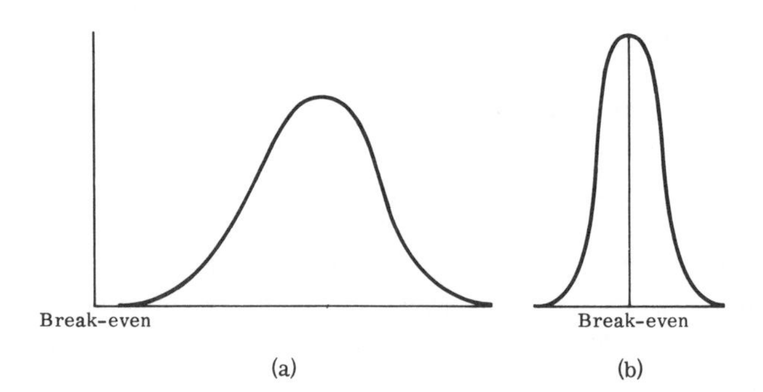

Figure 5.1 The effect of uncertainty relative to the break-even point. (a) Very uncertain, but far from break-even; little value to new information. (b) Little uncertainty, but near break-even; new information may be very valuable.

Example 5.1 (Continued) Political Campaigning: Effect of Uncertainty

The candidate's prior distribution, given earlier, was a possible 52% of the votes and the chances were two out of three that he would get between 50% and 54% of the vote. In this case, he might well seek additional information, since his prior expectation is uncomfortably near the break-even point of 50%. If, on the other hand, he expected to get 60% of the votes, there would be little value to new information, even if he were less certain. Thus, he might believe that the chances were two out of three that he would get between 55% and 65% of the vote (so that V = .05), but, since the probability of getting less than 50% is very small, he would be unwilling to spend much for more information.

5.7 LONG-RUN MEANING OF VALUE OF INFORMATION

It is impossible to guarantee that information will be valuable for every decision made. Suppose, for example, that an experienced decision maker is leaning toward a given course of action but is uncertain about the outcome, so he wishes additional information. In most cases, the information he receives will confirm the wisdom of his initial course of action. In these cases, except for the psychic benefit of reducing his anxiety, the information has been of no value. Information will have great value, however, in those cases in which it causes the decision maker to change his mind. Thus, if the mayor who wanted to start a new program is dissuaded from doing so by negative results, the city may save millions of dollars; on the other hand, positive results from a sample may persuade the city to start a project that otherwise would have been ignored.

Even without information, the capable decision maker will make the right choice, say, more than half the time. With information, he will increase the percentage of correct decisions. With perfect information, he will never make an incorrect decision, given that he behaves rationally and is never the victim of muddled thinking. It is unlikely, however, that perfect information will ever be available, since its cost would be more than its value. Thus, the value of information is the average increase in gains that occurs from making more right decisions and less wrong ones. Information increases the odds of being right, but it does not guarantee perfection.

It is possible to quantify the value of information if the decision maker's prior information or beliefs can be specified. Before doing this, let us consider the value of information in the case of basic research.

5.8 RESOURCE ALLOCATION BY PUBLIC AGENCIES FOR BASIC RESEARCH

How much should the federal government spend on cancer research next year, how much for welfare programs, how much for defense activities? These decisions require an evaluation of how valuable are cures for cancer, what is the value to society of reducing social ills, or, most fundamentally, what values can be placed on human life. Stated this baldly, many people would argue that it is impossible to answer these questions, yet these decisions are made continuously at both legislative and executive levels. To simplify the process, the decisions are made in many stages. First, overall decisions are made on what proportion of the gross national product to allocate to the economy's public sector. Then, within this sector, funds are allocated to major areas, such as defense and welfare, and the allocation process continues until funds are allocated at the lower levels to very specific projects. Even then, the directors of these projects must make decisions on fund allocations based on value judgments. Of course, the values of alternative decisions also depend on the state of the world: During periods of world tension, defense spending takes an increased portion of public funds; as a cure for cancer becomes increasingly probable, funds for cancer research are increased by transferring money from elsewhere.

Since it is so difficult to determine the value of decisions, mistakes are made frequently, or there are substantial disagreements about the value of certain decisions. Many public health officials felt that too much money was allocated for research in poliomyelitis during the 1950s because so small a portion of the population was affected, but the decision had been based on public opinion at the time. The value of landing men on the moon has also been questioned. What must be stressed is that decisions do get made even when there are disagreements about their value.

In questions of public policy, there are decisions that must be made even though the value of alternatives is determined by the political process. Is it possible to specify the value of basic research? Although this seems an even more difficult task, it is one that must be faced continuously by granting agencies like the National Science Foundation (NSF) and the National Institutes of Health (NIH). It is possible to solve this problem by substituting a notion of value or utility U for the monetary gains and losses of a decision. Given a fixed budget M, the funding agency selects projects to optimize U. The judgment about the total amount of funds M available to the agency is a political decision made through the legislative process.

In determining the value of a proposed project, the funding agency will consider three components comparable to those for the applied research decision: (*1*) the importance of the research, (*2*) uncertainty about the collected results, and (*3*) availability of prior research.

IMPORTANCE OF THE SUBJECT MATTER AREA OF THE RESEARCH TO THE FUNDING AGENCY

This first factor is expressed as a weight W_i for each project, with the ratio of the weights of two projects reflecting their relative importance. In the simplest case, a granting agency may decide merely whether or not the proposed project is within its scope, and assign equal weights of 1 to all acceptable projects and 0 to the rest. This procedure seems to be followed by university research boards, who are confronted by very little overlapping in the subject matter of projects.

A federal granting agency may assign measures of importance by ranking the projects on importance. After ranking, the agency or its review panel can assign importance weights to the highest- and lowest-ranking projects, with the intermediate projects getting intermediate weights. Funding agencies more concerned with applied research may assign a wide range of weights to different subject matter areas, ranging from, say, 100 for a very important project, to 5 for a project with only slight relevance for the mission of that particular agency. Funding agencies, such as NSF, that are concerned with more basic research would have a narrower range of weights ranging, perhaps, from 1 to 5.

The weight may also reflect political realities, such as the geographic location of the project, since funding agencies like NSF are required to give attention to the geographical distribution of funds among all 50 states. With geography as a factor, the proposals first would be sorted by state and region, and then, within each group, rankings and weights would be assigned.

UNCERTAINTY: THE VALUE OF A UNIT OF INFORMATION

Uncertainty for the ith project is expressed by σ_i^2. The value of a unit of information is the reciprocal of uncertainty, or $1/\sigma_i^2$, where σ_i^2 reflects the following factors: (*1*) the estimated variance in the population being studied, (*2*) measurement error, including judgments both about the reliability of the data and about biases or potential biases, and (*3*) the confidence that the funding agency has in the project director's ability.

Not all these factors may be present in any one project. The notion of variability of the population is a concept of the social, much more than of the physical, sciences. For example, a study of presidential election behavior, an event for which there are only two choices will have a smaller universe variance than will a study of income, expenditures, or savings. Comparing reports of current purchasing with anticipated purchasing, we will discover that the measurement error is lower for current purchasing. Thus, one unit of information teaches most about election behavior, less about current income, and least about anticipated expenditures. In a physical science application, where there is no universe variation, σ_i^2 reflects measurement error only. If both universe variation and measurement error are present, σ_i^2 is the sum of these variations, assuming that they are independent.

It is also observable that an organization like the Survey Research Center at the University of Michigan would be considered more capable of studying anticipated consumer expenditures, because of its long experience in this area, than would be a new Ph.D. from elsewhere. One way of expressing this confidence, or lack of confidence, is to assign a probability p between 0 and 1 so that the value of a unit of information is p/σ_i^2 or $1/(\sigma_i^2/p)$. Thus, the effect of lack of confidence in the project director's ability is equivalent to increasing σ_i^2. This can be considered similar to discounting the future expected income stream depending on uncertainty about the future.

We shall observe later that the value of a unit of information has an important effect on the level of partial funding, but not much effect on the decision to fund or not fund the project.

PRIOR INFORMATION

The more that is known about a topic, the less valuable new information will be. Thus, the funding agency usually would prefer to fund two projects partially rather than to fund one fully if costs, importance, and the value of a unit of information of the projects are equal. Prior information may be expressed by P_i, where $P_i = 1/V_i^2$, using the Bayesian notation for prior variance.

If the project reviewers know of extensive previous research on the same topic, there will be little remaining variance and P_i will be large. Assuming that information on sampling errors is available from previous studies, these could be combined in the usual way for independent samples, to give an estimate of V_i. As with the value of a unit of information, there could also be some discounting of previous information because of doubts about the methodology or analysis.

It may sometimes appear that funding agencies behave in a way just the opposite of the one suggested here. Funds are more likely to be available for a continuing project than for a new one, and the more information a project produces, the greater the likelihood of additional funds. This funding policy reflects the value of a unit of information: If there are useful results after the initial stage of a project, the confidence the funding agency has in the project increases substantially and thus increases the overall value of the project. Nevertheless, prior information is considered along with the other factors.

5.9 OPTIMUM SAMPLE SIZE FOR A BASIC RESEARCH PROJECT

The optimum sample size for a basic research project, along with the question of whether or not any sampling is done, is decided jointly by the researcher and granting agency. Usually, the researcher submits a proposal specifying a sample size. The granting agency must decide whether or not to fund the proposal at all; then, if the fund is granted, whether it should be a full or partial grant. Since the granting agency must consider competing projects, it frequently suggests a sample size smaller than that proposed by the researcher. The researcher then can accept or reject the partial funds.

For the granting agency, the choices of projects to fund and the level of funding depend on the fixed and variable costs of the competing projects, the amount of money available, and the value of information from the projects, as discussed earlier. The solution involves stratified sampling methods and is discussed in Chapter 6. Here, without giving the exact solution, we give some general implications of the optimum solution for the granting agency and the researcher.

IMPLICATIONS FOR THE GRANTING AGENCY

It will be seen in the next chapter that the question of whether or not a project i is funded at all is *not* primarily a function of sample size but of the importance W_i, the prior information P_i, and the fixed costs C_i. The effects of the fixed costs C_i are indirect. If the cost for one project is much above average, this prevents several other projects from being funded and leads to a decrease in the information available from these other projects.

For those projects that are funded, the key variables that determine sample size for the granting agency are $1/\sigma_i^2$, the value of a unit of information from the ith project, and c_i, the variable sampling cost. The problem of sample allocation among funded projects is identical to that of finding an optimum allocation of a stratified sample, but here the strata are the funded projects rather than subgroups of a population.

IMPLICATIONS FOR THE GRANT APPLICANT

If one takes this discussion seriously, it is evident that the grant applicant has an impossible task. The optimum sample size for one project depends not only on its cost and value but on the same factors for all competing projects. Given this dilemma, the director of any one project can only submit a maximum sample size n and anticipate that the final sample size is likely to be smaller. Nothing would prevent the granting agency from increasing the sample size if the maximum suggested n were too small, but this seldom occurs. More often, a sample size much too small indicates the grant applicant's lack of statistical competence.

The chief reason for asking for a large sample usually is that data are needed for a large number of subpopulations. The marginal gain from reducing the variance of the total population, given a reasonable sample size, is usually small relative to the gain in information from studying a subgroup.

Thus, assume a binomial variable, such as voting behavior, with p=.5. Then the standard error for a sample of 1000 is 1.6 percentage points, while for a sample of 2000, the standard error is 1.1 percentage points, a reduction of only .5 percentage points. On the other hand, consider a subgroup that is only 5% of the total population, such as women in a certain occupational category, or black men, or college graduates in a given region. Then the total sample of 1000 yields a sample of about 50 in this subgroup, and the standard error for the subgroup is 7.1 percentage points; a sample of 2000 yields 100 in the subgroup, and the standard error is 5.0 percentage points, or a reduction of 2.1 percentage points.

This strategy of asking for large samples to analyze subgroups cannot be extended indefinitely, since one ultimately is looking at very small fractions of the total population and the importance (W_i) of these groups becomes small.

It is my impression that, when grant applicants prepare their budgets, they give far more attention to variable costs than to fixed costs of data collection. Fixed costs include the project director's salary as well as costs for research assistants, equipment, and other overhead charges. The discussion here indicates, however, that fixed costs are more critical in determining whether a project is accepted or rejected, the variable costs being open to later negotiation. Thus, when in doubt, classifying a budget category as a variable cost should increase the project's likelihood of funding.

5.10 OPTIMUM SAMPLE SIZE FOR A TWO-ACTION LINEAR DECISION

For the special but important case in which the decision maker's prior distribution can be expressed as a normal distribution, with mean $\bar{x}_{\text{prior}}$ and variance V^2, it is possible to compute not only the value of information but the optimum sample size and the amount to spend on obtaining new sample information.

While a decision maker may seldom be able to specify precisely that his distribution is normal, he may frequently believe that it is symmetric around the most likely value and that the normal distribution is a reasonable approximation. Similarly, although it is difficult to specify initially a value for V, any two points on the normal distribution determine the parameters. Thus, if the decision maker believes that the most likely estimate of the number of persons who will benefit from a new service is 450,000, and he says the chances are two out of three that the number of persons will be between 350,000 and 550,000, then he has specified V, for he has specified the area to the right of the value U on the unit normal curve (see Figure 5.2).

From a table of the cumulative normal distribution, it may be seen that, if $P_N (U > u) = .167$, then u must $= .965$. Since u is defined as $u = |x - \bar{x}_{\text{prior}}| / V$, $.965 = 100{,}000/V$, and solving for V, $V = 100{,}000/.965 = 103{,}600$.

Both the value of information and the optimum sample size depend on how far the decision maker's value of $\bar{x}_{\text{prior}}$ differs from the break-even point $\bar{x}_b$ relative to the prior V. The distance from the break-even point D is defined as

$$D = \frac{|x_b - \bar{x}_{\text{prior}}|}{V}. \tag{5.1}$$

Schlaifer (71) has defined a function $G(D)$ as the unit normal loss integral which depends on the distance from break-even. As one would expect, this function has its highest value when $D = 0$, and becomes smaller as D increases. The values for the unit normal loss integral are given in Appendix C. It may be observed that the unit normal loss integral is a simple function of the height of the normal curve at D and the area of the normal curve for all values greater than D:

$$G(D) = P'_N(D) - DP_N(U > D), \tag{5.2}$$

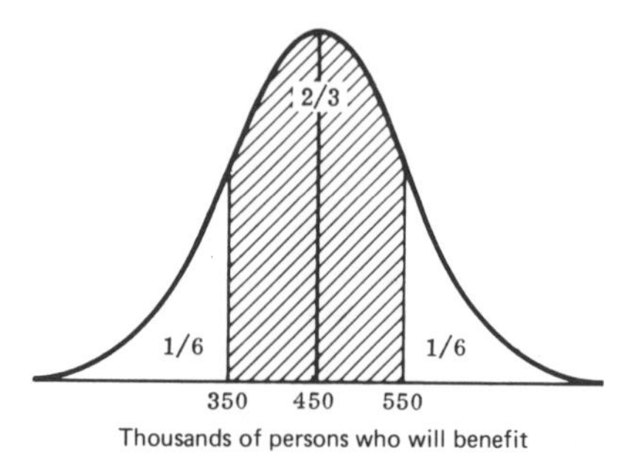

Figure 5.2 Bayesian normal prior distribution.

where $P'_N(D)$ is the height of the unit normal curve at point D, and $P_N(U > D)$ is the area under the unit normal curve to the right of point D.

The value of perfect information or the loss due to uncertainty (these are the same) is then found to be

$$L = k \cdot V \cdot G(D), \tag{5.3}$$

where k is the unit profit or social gain.

Example 5.2 Computing the Value of Information or the Loss Due to Uncertainty

Suppose that

$$P = -\$5{,}000{,}000 + \$10x.$$

Imagine, for example, that the \$5 million represents the cost of some program in a city, such as the building of a new park, swimming pool, or library. The gain of \$10 per person who uses the facility is estimated by comparing the costs with those of similar profit-making facilities. In the profit sector, the \$5 million would be the fixed costs required to start producing a new product, and the \$10 would be the unit profit per item sold. Then

$$x_b = \frac{5{,}000{,}000}{10} = 500{,}000 \text{ persons (units).}$$

If the decision maker's prior expectation is that 450,000 persons or units will use the new facility, and if his prior distribution is normally distributed with $V =$ 100,000, then he is relatively close to the break-even level for the project–further information to improve the estimate of the number of users will be of value. More precisely,

$$D = \frac{|500{,}000 - 450{,}000|}{100{,}000} = .5,$$

$$G(D) = .1978 \text{ from Appendix C,}$$

$$\text{and } L = (10)(100{,}000)(.1978) = \$197{,}800.$$

Although this amount is the maximum one would pay for perfect information, there would be no benefit in buying this much information because there would be no net gain from the reduction of uncertainty. The optimum sample size is the one for which the reduction of uncertainty $L-L'$ minus the cost of the information is maximized.

Suppose our cost function for sampling is the simple one developed in Chapter 4:

$$C = c_1 m + c_2 n.$$

Assume that $c_1 m$ is a fixed cost and that only n varies.

Then, Schlaifer has prepared a chart for finding the optimum sample size (reprinted here in Appendix D, p. 233). To use the chart, one must know the value of D as defined in Formula (5.1) as well as another statistic Z, where Z is computed from

$$Z = \frac{V}{\sigma} \sqrt[3]{\frac{k\sigma}{c_2}}. \tag{5.4}$$

The value of σ, the population standard deviation, must be estimated from previous studies of similar populations or from pilot test results. This is another example of the fact that optimum planning for a new study requires information from previous studies.

The values of Z and D on the chart determine a value of h. The optimum sample size is then found to be

$$n_{\text{opt.}} = h\left(\frac{k\sigma}{c_2}\right)^{2/3} \tag{5.5}$$

Example 5.2 (Continued) Computing the Optimum Sample Size

Suppose $\sigma = 1{,}000{,}000$

$c_2 = \$10$

Then $Z = \dfrac{100{,}000}{1{,}000{,}000} \sqrt[3]{\dfrac{10}{10} 1{,}000{,}000} = .1\ (100) = 10$

$D = .5$ (as before)

$h = .125$ (from Appendix D)

and $n_{\text{opt.}} = .125(100)^2 = 1{,}250.$

5.11 SHOULD THERE BE ANY SAMPLING AT ALL?

The solution method just presented gives the optimum sample size if there is to be any sampling at all, but does not indicate whether or not sampling is justified. That decision depends on the total cost of sampling and the reduction of uncertainty due to the optimum sampling. If the total costs of sampling exceed the anticipated reduction in loss due to uncertainty, no sampling is justified. To estimate the reduction in loss due to uncertainty, one must compute $L' = kV'G(D')$.

The value of V' is smaller than the value of V since there is now information from the sample. If one used only the information from the sample, the new value σ^2 would be $\sigma^2/n_{\text{opt.}}$. There is also the information available from the decision maker's prior distribution. To combine the sample information with the prior information, first note that the ratio $(\sigma/V)^2$ gives an equivalent sample size for the prior V. Thus, in Example 5.2, $(\sigma/V)^2 = 100$ means that the decision maker's prior information was the same as if a pilot sample of 100 cases had been surveyed. Then to find V', one combines the equivalent n from the prior distribution with the optimum n from the sample:

$$V' = \frac{\sigma}{\sqrt{\left(\frac{\sigma}{V}\right)^2 + n_{\text{opt}}}}. \tag{5.6}$$

If $L-L'-C$ is positive, sampling is justified and the difference represents the net gain due to sampling. Otherwise, no sampling should be conducted.

Example 5.2 (Continued) Should There Be Any Sampling?

Suppose

$$c_1 m = \$10{,}000.$$

Then the total cost of sampling 1250 cases will be \$22,500.

$$V' = \frac{1{,}000{,}000}{\sqrt{(10)^2 + 1250}} = 27{,}218$$

$$D' = \frac{50{,}000}{27{,}218} = 1.84$$

So $L' = 10 \cdot 27{,}218 \cdot G(1.84) = \$3511$, since $G(1.84) = .0129$ and $L-L' = \$197{,}800 - \$3{,}511 = \$194{,}289$. It is clear that, for this example, sampling is justified, and the net gain due to sampling is $\$194{,}289 - 22{,}500 = \$171{,}789$.

It is not true, in general, however, that sampling is always justified. We have already given examples of cases in which the decision maker had strong prior beliefs. In those cases, the relative decrease in uncertainty is small; and, if there are large fixed costs in sampling, these fixed costs often can be larger than the decrease in uncertainty.

There are additional costs, in terms of time delay and foregone benefits, that need to be considered since they are sometimes much larger than the direct sampling costs. Thus, in marketing, one often sees examples of new products that are rushed to the market without any field testing or survey work. The fact that most of these products are failures does not mean the decision makers have behaved irrationally. When questioned about the decision, the typical response is that the product was rushed to market to anticipate the competition who were developing similar products. Here, a delay of 6 months to 1 year for survey work would result in substantial lost profits if the product were a success. On the other hand, many new products are sufficiently like existing products, or sufficiently unique, that the time delay need not be considered. In general, any firm that *always* seeks new sample information or *never* seeks such information is not optimizing profits.

In the nonprofit sector the same nonmonetary factors must be considered. Policy decisions must often be made without sufficient time for research. On the other hand, a common bureaucratic procedure for avoiding or postponing a necessary decision is to seek additional information, even if there already is available extensive information. A major objection of many community groups in poor neighborhoods is that research is used to replace rather than to guide action programs. As in the profit sector, however, rushing into action programs without adequate research leads to wheel-spinning and frequently to funds being mostly wasted.

Individual basic researchers also must be concerned about time, especially early in their careers. An important project sometimes must be postponed or discarded because it would take several years to gather the data. Since academic promotion and the availability of resources for additional research usually depend on evidence of professional publications, the tendency is to work on secondary data or on small studies for which the data can be gathered quickly. Senior researchers with tenure and with some backlog of data to analyze while waiting for the new results are more likely to undertake long-term panel studies.

5.12 SUMMARY

The determination of sample size is a difficult task requiring both estimation of the various costs related to sampling (including time) and of the value of the new information. The value of information is a subjective decision depending on (*1*) the importance of the research or the gains and losses associated with alternative decisions; (*2*) what is already known, either from earlier research or from less formal experience of the decision maker; and (*3*) the effect of uncertainty on the decision process.

More generally, sample size determination is an example of the allocation of scarce resources to optimize some value function. For basic research, it is obvious that this decision is made ultimately by the granting agency, not the grant applicant.

Tables 5.1 and 5.2 at the beginning of the chapter give examples of sample sizes currently used by social scientists. Although these examples may be of aid in initial planning, they depend on current costs and data-collection methods and cannot be used blindly.

5.13 ADDITIONAL READING

The reader who is completely unfamiliar with Bayesian statistics and wishes to learn more will be interested in Kyburg and Smokler's reader, *Studies in Subjective Probability* (42). Most of the articles are philosophical and expository and do not require a strong mathematical background. Two other important books that do require mathematical sophistication are Savage's *The Foundations of Statistics* (70) and Raiffa and Schlaifer's *Applied Statistical Decision Theory* (66).

Some psychological research uses of Bayesian statistics are found in papers by Edwards (20, 21) and in a paper by Edwards, Lindman, and Savage on "Bayesian statistical inference for psychological research" (22). Novick and his collaborators (58, 59, 60) at the American College Testing Program have been using Bayesian procedures in educational research. An excellent discussion of Bayesian procedures in econometrics is given by Zellner (102), but this book requires a strong mathematics background.

The discussion of the decision framework for a granting agency is taken from Ericson (23). This discussion is continued in the next chapter on stratified sampling. The discussion of the two-action linear decision problem is based on Schlaifer's *Probability and Statistics for Business Decisions* (71, Ch. 30, 35). The best non-Bayesian discussion of sample size determination is by Cochran (14, Ch. 4).

6

Stratified Sampling

It is frequently useful to divide the total population into subgroups called "strata" for purposes of making the sample more efficient. In this chapter, we discuss when stratified sampling is and is not appropriate and how to determine optimum strata sample sizes.

In this chapter, we assume that the amount of money to be spent on data collection has already been determined, and that the methods and costs of the data collection are known. Given these constraints, the efficiency of a sample is measured by the size of its sampling error relative to other samples that cost the same. Stratified sampling is intended to provide the smallest sampling error and hence the most information for the available resources.

It will soon be clear to the reader that there is no optimum all-purpose stratified sample. Different designs are optimum depending on the topics studied and how the data are to be analyzed. The most troublesome problem is deciding on a sampling strategy when a single study is used for multiple purposes; it is seldom possible to find a sample that is optimum for all purposes. Instead, some compromise sample must be found. This is discussed later.

There are four principal situations for the use of optimum stratified sampling:

1. The strata are themselves of primary interest.
2. Variances differ between the strata.
3. Costs differ by strata.
4. Prior information differs by strata.

Sections 6.2–6.5 discuss these situations. The final section describes estimation procedures and sampling error computations for stratified samples.

6.1 APPROPRIATE AND INAPPROPRIATE USES OF STRATIFICATION

Stratified sampling is frequently considered by researchers who basically have no belief that probability sampling methods work. Used with probability sampling, these procedures can sometimes greatly increase field costs without improving the sample. Used instead of probability sampling procedures, the sample may actually be far worse than a probability sample. Before turning to appropriate uses of stratification, it may be worthwhile to warn of inappropriate uses.

Example 6.1 Inappropriate Use of Stratification to Ensure Randomness

Naive researchers with high anxiety levels will often be concerned that a sample using probability methods will yield very strange results, a large oversampling of men, an undersampling of middle-aged persons, distorted income distributions, or other catastrophes. To ensure against this, they insist that the sample be stratified so that age, race, sex, occupation, income, education, household size, and other variables are guaranteed to be perfectly represented. (Lest any reader think I am setting up a straw horse to be demolished, this example is based on numerous requests for just such sampling.) If the request is taken literally, there is simply no way to comply with it. The census data that would be necessary for strata controls would not be available in such detail, to preserve the privacy of individual households. Even if strata totals could be estimated, the cost of perfectly matching these in the field would be prohibitive since hundreds of additional screening interviews would be required to match strata controls exactly. Finally, the effects on the data of this tortuous process would probably be undetectable.

Example 6.2 Inappropriate Use of Stratification with Nonprobability Samples

Before area probability sampling methods were developed, a procedure frequently used was the establishing of quotas for the interviewer. The interviewer was requested to get five men and five women: one woman and one man with "high" incomes, two of each with "middle" incomes, and one of each with "low" income. Such instructions, or those specifying other demographic controls, never specified precisely where or how the sampling was to be done, so the interviewer typically found respondents in the most convenient locations, generally those close to her home. Not only did these procedures introduce serious biases, but the idea of a known sampling variability became almost meaningless. Estimates of sampling errors depend on the assumption that it is possible to draw repeated samples, using the same procedures. When each interviewer can do as she chooses, and can change procedures from one study to another, no real replication is possible. The use of strata quotas did not correct for these defects but merely concealed them for the most obvious variables.

Today, very little geographically uncontrolled sampling is done, except for pilot testing. Here the use of strata controls can be helpful in pinpointing subgroups who may have special difficulty with the questionnaire. (Many survey organizations use a form of quota sampling rather than repeated callbacks to account for respondents who are not home. This form of quota sampling has well-specified characteristics and can be replicated. It is discussed in Chapter 9.)

Example 6.3 Inappropriate Use of Stratification to Adjust for Noncooperation

Another spurious use of stratification is to correct for major problems in sample cooperation. Even if the sample design is excellent, little trust can be put in a sample in which only 5% or 10% of the population cooperate. (Such low cooperation is often found in poorly designed mail surveys of the general population.) Then, the hapless data collector will often attempt to demonstrate that the sample is adequate by comparing the sample distribution to known census data for several demographic characteristics. Even if there are no large biases in these comparisons, there is still no assurance of the quality of the data for unmeasured variables. Usually, with very low cooperation, there will also be some noticeable biases in the demographic characteristics. Then, as an effort to salvage something from the wreck, the data are weighted on one or more demographic variables to correct for the sample biases. The weighting is likely to correct only for the variables included, not for the major variables for which information is being collected.

The use of weights to correct for sample biases due to noncooperation is called "post-stratification," because it is done after the sampling. It is a legitimate procedure to correct for minor differences between the sample and the universe due to sampling error and differential cooperation. Thus, in national surveys, post-stratification is sometimes used to adjust for the fact that cooperation is lower in central cities than in rural areas. Except for very large studies, such as that in Example 6.4, from which very precise national estimates are required, post-stratification is usually not worth the added complexity of data handling and analysis.

The reader may wonder, after the last several examples, whether or not it is ever appropriate to stratify a sample if the strata are to be sampled proportionately. Under some circumstances, stratification will be inexpensive and some gains can be expected. This occurs when the stratification is of primary sampling areas (PSUs) for a national survey rather than of households. Stratification of PSUs is done during the normal sample selection process and need not require any additional field costs. Usually, a systematic sample of PSUs is selected by using the multistage sampling procedures described in Chapter 7. In this case, the stratification involves arranging the PSUs in order by variables that are expected to be related to the variables that are to be studied, rather than arranging the PSUs in alphabetic order or in a haphazard fashion. The variables usually used are region of the country, degree of urbanization, proportion of nonwhite, and some socioeconomic measure, such as income or education.

The effects of such stratification are moderate. Sampling errors are generally less than 10% smaller than they would have been without stratification; but, unless much time and expense will be wasted in agonizing about how to stratify, there is no reason not to do so.

Example 6.4 The Current Population Survey

The CPS, a very large and careful sample, has already been discussed in Chapter 1. The PSU stratification involves the following variables:

Standard Metropolitan Statistical Area or not
Rate of population change
Percentage of population living in urban areas
Percentage of population in manufacturing
Principal industries
Average value of retail trade
Proportion of nonwhite population

A trenchant comment on the stratification efforts is found in a technical paper prepared by the Census Bureau on the methodology of the CPS (92):

> A great many professional man-hours were spent in the stratification process. However, it is questionable whether the amount of time devoted to reviews and refinements paid off in appreciable reductions in sampling variances. Intuitive notions about the gains from stratification can be misleading. Methods of stratification that appear to be very different often lead to about the same variances [p. 6].

6.2 THE STRATA ARE OF PRIMARY INTEREST

Seldom do social scientists look only at the total population. Even media articles, such as results from the Gallup Poll, repor on attitudes correlated with race, sex, education, region of the country, and other demographic variables. For most Gallup Polls, however, it is evident that reports about the total population are more important than information about subgroups.

Suppose, on the other hand, one is concerned about the effects of discrimination on wages paid to minority groups, such as women and blacks. Here, total population figures are of little interest. The goal is to compare the incomes of blacks and whites, women and men, minority and majority groups.

The optimum sampling designs for these two cases can be quite different. For the case in which estimates of the total population are of primary importance, one needs a proportionate or self-weighting sample of the entire population. *For comparison of subgroups, the optimum sample is one where the sample sizes of the subgroups are equal, since this minimizes the standard error of the difference.*

If one wishes to compare men and women on any given variable, as well as make estimates of the total population, there is no problem, for the two groups are almost evenly split in the population. If, however, one wishes to compare blacks and whites, a proportional sample will be very inefficient because nine-tenths of the sample will be white, and one needs a sample with equal numbers of whites and blacks. For this case, one would simply increase the black population sample by about nine times the rate of the white population to get equal samples. In general, when comparing two groups, the less common group should be sampled at the rate p/q relative to the more common group, where p is the proportion of the more common group and q of the less common group. For three or more groups, the sampling rates for any two of the groups are inversely proportional to their proportions in the populations. Thus, if Group A is 60% of the total population, Group B 30%, and Group C 10%, the sampling rate for Group C is six times the sampling rate for Group A and three times the sampling rate for Group B, while the sampling rate for Group B is twice that for Group A.

Unfortunately, when the sample is optimized for group comparisons, it is no longer optimum for estimating the total population. This may not be of concern if one is uninterested in totals, but the most common situation is to be interested in both. Three strategies are used in this case, all compromises between the different needs.

1. *Split the difference.* Suppose one were equally concerned about total unemployment in a community and the unemployment rates of blacks and whites. A sample of 30% black and 70% white, or one-third black and two-thirds white, could be a reasonable compromise between the extremes of proportional and equal-sized samples.
2. *Determine a minimum acceptable total sample size, if sampling is proportional, and use remaining resources to augment the smaller subgroups.*
3. *Determine minimum subgroup sample sizes if equal-sized samples are se-*

lected for each subgroup, and use remaining resources to improve estimates of the total.

Since we saw in the last chapter the problems involved in determining minimum sample sizes, the "split the difference" procedure is probably the most widely used.

If disproportionate sampling is used, it will normally require screening the population before the main data-gathering interview. That is, an initial probability sample must be selected which is large enough to include the required number of respondents for the final sample. All elements in the initial sample receive a classifying interview. All the respondents in the most oversampled stratum identified in this first interview receive the main interview, and respondents in the other strata are subsampled.

Example 6.5 Studies of Unemployment and Underemployment in Poverty Areas

A government agency wishes to conduct a study in a specified city neighborhood, comparing the work histories, educational backgrounds, and attitudes toward training programs of persons in the labor force who are currently unemployed or underemployed to those who are currently employed. They wish to get samples of 500 in each group.

The total population consists of 100,000 persons in the neighborhood who are in the labor force, of whom it is estimated that 75,000 are employed and 25,000 are unemployed or underemployed. Both the groups are spread evenly throughout the area and can be located only by screening.

Ignoring noncooperation for this example, the agency selects an initial screening rate of 25,000/500, or 1 in every 50. This will yield an initial sample of 2000 respondents, of whom 1500 are employed and 500 unemployed or underemployed. All of the latter group are interviewed during the main study, and a random one-third of the employed group get the final interview.

6.3 VARIANCES DIFFER BETWEEN STRATA

The earliest use of optimum stratified sampling was by Neyman (57) who, in his classic 1934 paper, demonstrated that, if strata variances differ, the optimum sample allocation among strata is given by

$$n_h{}^* = \frac{N_h \sigma_h}{\sum_h N_h \sigma_h} n$$

or

$$n_h{}^* = \frac{\pi_h \sigma_h}{\sum_h \pi_h \sigma_h} n \qquad (6.1)$$

where

N_h = total elements in the population in Stratum h
π_h = proportion of total population in Stratum h
σ_h = standard deviation in Stratum h
n = the total sample size
$n_h{}^*$ = optimum sample size in Stratum h

The principal use of Formula (6.1) is *not* with human populations but with institutional populations, such as universities, schools systems, and hospitals, and with business firms. We shall see that the differences in variances between large and small firms and institutions are far greater than between persons, for most variables. Optimum sampling procedures for human populations are sometimes used on expenditure studies, but this requires an initial screening. Example 6.6 indicates why disproportionate sampling is not normally used on attitude studies for human populations. The next example illustrates optimum sampling of households in a medical expenditure study. The last example in this section (6.8) shows the large increase in sample efficiency that results from optimum sampling of hospitals.

Example 6.6 Why Optimum Sampling Is Not Used on Attitude Studies (Hypothetical)

Suppose one wished to study attitudes about the current U.S. foreign policy toward China. Based on previous studies, it could be expected that about 70% of Democrats, 50% of independents, and 20% of Republicans favor the current policy. Since the differences between the groups are so large, one might consider disproportionate sampling of the three. Assume that the population consists of 40% Democrats, 35% Republicans, and 25% independents, so a proportionate sample of 1000 respondents would include 400 Democrats, 350 Republicans, and 250 independents. To find the optimum sample, it would be necessary to compute the standard deviations (σ_h) which here are found from computing $(PQ)^{1/2}$ (see Table 6.1).

It may be seen that the differences between the proportional and optimum samples are so small that it would not be worth the increased complexity of

Table 6.1
Computations for Optimum Sample for Study of Attitudes on Foreign Policy

Stratum	π_h	P_h	$\sigma_h = (P_h Q_h)^{1/2}$		$\pi_h \sigma_h$	$\frac{\pi_h \sigma_h}{\sum_h \pi_h \sigma_h}$ 1000
Democrats	.40	.7	.46		.184	410
Republicans	.35	.2	.40		.140	312
Independents	.25	.5	.50		.125	278
				$\sum_h \pi_h \sigma_h =$	.449	1000

screening the population and of data handling to improve the estimate a very little bit. The differences are small because the σ_h are not very different, although the P_h are farther apart than would usually be found in reality.

Example 6.7 Optimum Sampling for a Study of Health Expenditures

If one is attempting to measure total health expenditures of all households in the United States, a proportionate sample in which all households have the same probability of selection is not optimum. It is known from previous studies (2, 3) that the variance σ^2 in expenditures is more than 20 times greater for households with high expenditures than for the other households. An optimum design would be to screen a larger probability sample and then oversample those with high expenditures. Table 6.2 gives the computations.

If a total sample of 2000 cases is desired, this would mean a sample of 1120 low-expenditure households and 880 high-expenditure households. Again, ignoring noncooperation, it would be necessary to screen 880/.15, or about 5870 households, to get 880 high-expenditure households. To get 1120 from the remaining 4990 households would require subsampling and selecting every 4.5th household, or 2 of 9.

It is desirable to conduct the major interview, if one is required, along with the screening interview. To do this, it is necessary to establish in advance the criteria for determining high-expenditure households. For the example given, the definition of a high-expenditure household was one in which the family reported having spent more than $500 for medical care in the survey year (1958) or reported the death of at least one family member in the same period. In subsequent studies, the same principle has been used, but the amount of expenditures has increased, due to inflation and increased medical costs.

When this study was conducted, the optimum sampling rate for high-expenditure households was not used because the study had multiple purposes. In addition to information about expenditures, the survey also attempted to obtain information about insurance coverage. For estimating the proportion of respondents with insurance coverage, a self-weighting proportionate sample was optimum. As suggested in the last section, a compromise sample was actually designed which consisted of about 30% high-expenditure households, splitting the difference between 15% and 44%.

Table 6.2
Computations for Optimum Sample for Study of Health Expenditures

Stratum	π_h	σ_h^2	σ_h	$\pi_h \sigma_h$	$\pi_h \sigma_h / \sum_h \pi_h \sigma_h$
High expenditures	.15	$20\sigma^2$	4.5σ	$.68\sigma$	.44
Low expenditures	.85	σ^2	σ	$.85\sigma$	.56
				$\sum_h \pi_h \sigma_h = 1.53\sigma$	

OPTIMUM SAMPLING OF INSTITUTIONS

It is a general (although not universal) rule that, as elements become larger and more complex, the variability between them increases. Thus, the difference in the number of patients served or in the total annual budgets is far greater between the 10 largest hospitals than between 10 small hospitals. This suggests that institutions and firms be stratified by size and that larger samples be taken from the strata consisting of the larger-sized institutions. If the variances for each stratum are known from past experience, Formula (6.1) is used directly as in Example 6.7.

If the variances are not known, a measure of size for the strata may be used to approximate the variances, as in Example 6.8. Size measures may be based on the number of employees, the number of clients, the annual budget, the total sales, the amount of equipment, the square feet of space occupied, or other size variables. If more than one size measure is available, the one most closely related to the study's critical variables should be used. In a study of personnel policies, the number of employees would be a better size measure than sales or the number of clients served. This measure of size will not usually be perfectly correlated with the variances, so the sample rates chosen will not be quite optimum but still will be far more efficient than a proportionate self-weighting sample.

Example 6.8 Optimum Sampling for a Study of Hospitals

Suppose one is interested in the employment practices of hospitals, wage rates, employee benefits, treatment of minority employees, recruitment procedures, and so on. The complete listing of hospitals is found in the Guide issue of *Hospitals,* published by the American Hospital Association. Table 6.3 gives data on the payroll and the number of employees for hospitals broken into six size groups. (The data are based on a sample of the 1971 list.)

Table 6.3
Characteristics of U.S. Hospitals, by Number of Beds

Number of beds	Number of hospitals	π_h	Average payroll	Payroll σ_h	Average number of employees	Employees σ_h
Under 50	1614	.246	266	183	54	25
50–99	1566	.238	384	316	123	51
100–199	1419	.216	1,484	641	262	95
200–299	683	.104	3,110	1347	538	152
300–499	679	.103	5,758	2463	912	384
500 and over	609	.093	10,964	7227	1548	826
Totals	6570	1.000				

Table 6.4 gives the sample allocations under five different procedures for a total sample of 1000 hospitals. Estimates of the total sampling variance are also given for the payroll and the number of employees. The formulas used for computing these variances are given in Section 6.7 of this chapter, where this example is continued.

1. The proportional sample allocates the sample based on the number of hospitals per stratum, or by π_h.
2. The optimum payroll sample allocates on the basis of

$$n_h = \frac{\pi_h \sigma_h \text{ (payroll)}}{\sum_h \pi_h \ \sigma_h \text{ (payroll)}} n.$$

3. The optimum employees sample allocates on the basis of

$$n_h = \frac{\pi_h \sigma_h \text{ (employees)}}{\sum_h \pi_h \ \sigma_h \text{ (employees)}} n.$$

4. Estimates (1) and (2) in Table 6.4, using total beds, assume that only the data on the number of hospitals in each stratum are available. It is still possible to estimate size by taking the midpoint of the interval and multiplying by the number of hospitals. Thus, the number of beds in hospitals with under 50 beds is estimated by multiplying the midpoint 25 by 1614 for an estimate of 40,350. Similarly, in the stratum for hospitals with 300–499 beds, the estimated number of beds is 400 X 679, or 271,600. It is often the case, as in Table 6.3, that there is no way of determining the midpoint for the largest stratum. Estimate (1) assumes that the midpoint is twice as large as the lower bound, that is, 1000 beds.

Table 6.4
U.S. Hospital Study Sample Sizes Using Various Procedures and Total σ^2 for n=1000

Number of beds	Proportional	Optimum: Payroll	Optimum: Employees	Using total beds as measure of size: (1)	Using total beds as measure of size: (2)
Under 50	246	34	36	28	19
50–99	238	57	71	83	56
100–199	216	104	120	150	103
200–299	104	106	93	120	82
300–499	103	192	231	191	131
500 and over	93	507	449	428	609
Totals	1000	1000	1000	1000	1000
σ^2 employees	71.0	17.1	16.5	17.2	20.6
σ^2 payroll ('000)	4908	871	908	941	982

Estimate (2) avoids the issue by taking all the hospitals in the largest stratum and allocating the remaining sample of 391 hospitals among the remaining strata. The allocation is simply on the basis of size and requires no previous estimate of σ. (In many real-world applications, we would have size measures but no estimates of σ.)

$$n_h = \frac{s_h}{\Sigma s_h} n$$

where s_h is the estimated number of beds in the stratum.

The important feature of Table 6.4 is that all the size allocations are similar and all result in variances only one-fourth or one-fifth as large as those from proportional sampling. To put it another way, a proportional sample would need to be four or five times larger to have the same variance as that of the sample using size measures.

Which of the samples using size measures is the optimum? The table indicates that no sample is optimum for all purposes. Obviously, if one were interested only in payroll data for given types of workers, the optimum payroll sample would be best. If one were interested in recruitment procedures, the optimum employee sample would be best. But the differences between these two samples are slight: Using even the number of beds would increase the variance only a little for these variables and number of beds might be optimum for others, such as studies of patient care. If multiple measures of size are available, one should choose the one that is most current and that appears to be most highly correlated with the study's most important variables. As the example illustrates, however, it is not necessary to agonize over the choice or to search for a perfect measure to achieve major improvements in sample efficiency.

SAMPLING ALL ELEMENTS IN THE LARGEST STRATUM

In many populations, the number of institutions in the largest stratum will be small, but the variance in this stratum will be very large. This will be true especially for business firms that have multiple plants or retail establishments. The variance in the largest stratum will be much larger than for hospitals, which are limited to a single geographic location. For these populations, it will be the case that any optimum allocation using a size measure would indicate a sample size in the largest stratum that is larger than the number of elements in that stratum. The simple solution is to take *all* the elements in the largest stratum. If this is done, there is then *no* sampling variance in that stratum. This alone usually reduces substantially the total sampling error. The total sample remaining is then allocated among the remaining strata, as in the previous example.

Example 6.8 (Continued)

Suppose that, instead of a sample of 1000 hospitals, the sample size is 1500. Then Table 6.5 gives the optimum allocations. Note that, in all cases, the

Table 6.5
U.S. Hospital Study Sample Sizes for n=1500

Number of beds	Optimum		Using total beds as measure of size
	Payroll	Employees	
Under 50	61	58	44
50–99	103	115	129
100–199	189	194	233
200–299	191	150	187
300–499	347	374	298
500 and over	609	609	609
Totals	1500	1500	1500

largest stratum is completely sampled, there is only one estimate using total beds as a measure of size, and all three procedures give very similar optimum samples.

6.4 COSTS DIFFER BY STRATA

In Chapters 1 and 3, we discussed the use of combined methods for improving sample efficiency. The major feature of these combined methods is that the costs vary, depending on the procedure used. Thus, using phone interviews in combination with personal interviews for those without phones, the costs are much higher in the nonphone stratum than in the phone stratum. Similarly, if the procedure involves mail questionnaires followed by telephone calls to non-respondents, the costs will be higher for the phone stratum than for the mail stratum.

An optimum procedure in this case, as shown by Neyman (57) and by Hansen, Hurwitz, and Madow (35), is to sample strata inversely proportional to the square roots of per-interview costs c_h in the various strata. Combining this result with the one in the previous section, the optimum sample allocation by strata is

$$n_h{}^* = \frac{\pi_h \sigma_h / \sqrt{c_h}}{\sum_h (\pi_h \sigma_h / \sqrt{c_h})} n. \tag{6.2}$$

Unike the examples in the last section, most applications for which costs differ deal with human populations, so usually the variances are equal in the strata and the formula simplifies to

$$n_h{}^* = \frac{\pi_h / \sqrt{c_h}}{\Sigma(\pi_h / \sqrt{c_h})} n. \tag{6.3}$$

Since the sampling rates depend on square roots of costs, small differences like those observed in differing regions or city sizes can be ignored, but major differences in cost by type of procedure can substantially influence sample allocations.

Example 6.9 Combined Use of Phone and Face-to-Face Procedures

A researcher has a budget of \$10,000 for collecting data on political participation in a metropolitan area. He estimates that about 80% of all households have telephones. The costs of telephone interviewing are estimated at \$9 per completed case, based on previous experience with interviews of the same length. The costs of face-to-face interviewing are \$25 if all interviews are conducted this way, and \$36 if only nonphone households are interviewed face-to-face. The difference is a result of the additional travel and screening costs of locating nontelephone households. Assuming the variances in the two strata are similar, Table 6.6 gives the optimum sampling design, taking costs into account, and compares it with a sample using all face-to-face interviews or with one in which the two strata are sampled proportionately. (The use of telephones only is not considered because of the concern about sample biases if those without phones are omitted.)

It may be seen that the optimum sample has a variance of only about half that resulting from the use of all face-to-face interviewing, even though screening is required in the combined procedure. The optimum sample variance is about 10% lower than that of proportional sampling. This is achieved by conducting an additional 188 phone interviews rather than 47 more face-to-face interviews. Note that the reductions in variance due to costs per stratum are important but are generally smaller than those in the previous section.

It may be helpful to indicate how one computes the total sample size n given a fixed budget:

$$n = \frac{\text{Budget}}{\sum_h p_h c_h}$$

where p_h is the proportion of the sample in the hth stratum.

TABLE 6.6

Sample Sizes Using Various Procedures and Total σ^2 for \$10,000 Budget

		All face-to-face		Combined methods			
Stratum	π_h	c_h	Proportional	c_h	Proportional	$\pi_h/\sqrt{c_h}$	n_h^*
Phone	.80	\$25	320	\$ 9	555	.267	743
No phone	.20	25	80	36	139	.033	92
Totals	1.00		400		694	.300	835
σ^2			σ^2		$.58\sigma^2$		$.52\sigma^2$

$$P_h = \frac{\pi_h/\sqrt{c_h}}{\Sigma(\pi_h/\sqrt{c_h})}.$$

Thus, for the proportional sample

$$n = \frac{10{,}000}{.8(9) + .2\,(36)} = \frac{10{,}000}{14.4} = 694.$$

For the optimum sample, 89% should be phone interviews (.267/.300) and 11% face-to-face. Here

$$n = \frac{10{,}000}{.89(9) + .11(36)} = 835.$$

Example 6.10 Combined Use of Mail Procedure and Personal Follow-Up (Example 1.11 Continued)

A study was conducted to determine the current attitudes of physicians toward smoking and their own smoking behavior. The results were used by the Public Health Service in anti-cigarette-smoking campaigns. A budget of \$50,000 was allocated for data collection. It was estimated that about 40% of the physicians would respond to a mail survey with two follow-ups, at a cost of about \$1 per completed case; the other 60% would require either a long-distance phone interview or a face-to-face interview, at a cost of about \$25 per completed case. Table 6.7 gives the optimum sampling design and compares it to less desirable alternatives. Here, the variance for the optimum design is 58% lower than for proportional sampling, illustrating that the advantages of optimum allocation increase as the cost differences become larger.

The initial mailing to physicians required a mail sample of 14,762 to yield the optimum sample size of 5905, given that only 40% of the mail sample would respond. The cost of \$1 per completed case includes the cost of the total mailing. From Formula (6.3), the sample of 1764 doctors to be contacted personally is one-fifth of the initial sample who did not cooperate, since $\sqrt{\$1/\$25} = 1/5$. The least desirable alternative is not shown in Table 6.7 although it is common among naive researchers—the use of mail only. Unless one

Table 6.7
Sample Sizes for Study of Physicians' Smoking Habits

		All personal		Combined methods		
Stratum	π_h	c_h	Proportional	c_h	Proportional	Optimum
Mail responders	.4	25	800	1	1299	5905
Personal responders	.6	25	1200	25	1948	1764
Totals			2000		3247	7669
σ^2			σ^2		$.6\sigma^2$	$.46\sigma^2$

expects mail response to be as high as can be achieved by personal methods, funds must be left in the budget for the personal follow-ups.

6.5 PRIOR INFORMATION DIFFERS BY STRATA

In the last chapter, we discussed the use of Bayesian analysis for determining optimum sample sizes. Readers who skipped that section or had substantial difficulty with it are well-advised to skip Sections 6.5 and 6.6, at least during the first reading. Those brave souls who continue, however, will find that, even though the formulas look foreboding, they are really not especially difficult to understand or to use. Ericson (24) has shown how Bayesian analysis may be used to determine optimum stratified samples when some information is known already. In a sense, this is merely a generalization of a notion of the combined procedures discussed in the last section, except that some of the data are obtained from the prior experience of the researcher or decision maker. If, for example, the researcher has had professional experience and contacts with certain hospitals, he may decide to exclude these from further sampling. Similarly, a business firm may be in constant contact with its largest customers and feel no need for further sampling from this group when researching a new product, although the variance in this stratum is much larger than among smaller customers.

Although it is often difficult to specify differential prior knowledge by strata, it is often necessary to do so, as in the selection of basic research projects, discussed in Chapter 5. This is an example of choosing an optimum stratified sample when both fixed and variable costs of sampling differ by strata. The problem is somewhat less complex if only variable costs differ by strata. Examples of both cases are given here. Using the previous notation,

C = total funds available for data collection
π_i = proportion of the population in the ith stratum
c_i = cost/case in the ith stratum
σ_i^2 = population variance in the ith stratum
V_i^2 = the decision maker's prior variance in the ith stratum.

If B_i is defined as $\sigma_i\sqrt{c_i}/\pi_i V_i^2$ and the strata are numbered so that $B_o > B_1 > B_2 \ldots B_k$, where $B_o = \infty$, then the solution consists in sampling the subset of strata so that the optimum sample sizes are

$$n_i = 0 \qquad i \leqslant r$$

$$n_i = \frac{\pi_i \sigma_i}{\sqrt{c_i}} \left[\frac{C + \sum_{j=r+1}^{L} (c_j \sigma_j^2 / V_j^2)}{\sum_{j=r+1}^{L} \pi_j \sigma_j \sqrt{c_j}} \right] - \frac{\sigma_i^2}{V_i^2} \qquad i > r \tag{6.4}$$

where $r = 0,1,2, \ldots k-1$ is found by

$$I_r = \{C | C_r \leqslant C \leqslant C_{r-1}\}$$

and

$$C_r = B_{r+1} \sum_{i=r+1}^{L} \pi_1 \sigma_1 \sqrt{c_i} - \sum_{i=r+1}^{L} \frac{c_i \sigma_i^2}{V_i^2}$$

$$C_{-1} = \infty$$

Although Formula (6.4) looks formidable, there are really only two differences between it and Formula (6.2), which ignores prior information. The final term σ_i^2 / V_i^2 indicates, as we have seen in Chapter 5, the equivalent sample size information already available for the ith stratum. Perhaps the even more significant difference is that the availability of some prior information about a stratum makes it possible that *some strata will not be sampled at all.* As the next example illustrates, Formula (6.4) explains why optimum sampling designs do not require 100% cooperation.

Example 6.11 Optimum Sampling for Nonresponse

Suppose we use the same problem as that in Example 6.7, the selection of an optimum sample to measure medical care expenditures, but now primarily concern ourselves with the problem of nonresponse. We know from past experience that there are three main classes of respondents. The easy 60%, such as unemployed housewives, retired people, or households in rural areas, cost $25 per case to complete; the more difficult 20% of the cases, such as employed men or households in large cities, cost $36 per completed case. The most difficult group of respondents who are typically missed in surveys are not *truly* impossible to interview. Given sufficient effort and some compensation to the respondent, it should be possible to reach even the difficult, but the per-case cost would be much higher, possibly several hundred dollars or more. To be specific, let us assume an average cost of $900.

From previous studies, by comparing the completed sample to outside validating data from hospitals and insurance companies, we feel reasonably confident that the medical expenditures of this "most difficult" stratum are about the same as those of the other two strata. We also feel that we can make estimates about medical costs, based on earlier studies' results updated by using the Consumer Price Index, but that these estimates are crude and must be refined using new sample information. A budget of $35,000 is available for new data collection.

Table 6.8 gives the values of σ_i^2 and V_i^2 and other computations required for determining the optimum sample. Note that, in each case, the ratio of σ_i^2 / V_i^2 is 50, indicating that our prior information is equivalent to a sample of 50 in each stratum. The value of σ_i^2 is higher in the easy stratum because medical expenses increase for retired persons.

Table 6.8
Computation of Optimum Sample with Prior Information[a]

Stratum	π_i	c_i	$\sqrt{c_i}$	σ_i^2	V_i^2	σ_i	B_i	$\frac{\pi_i\sigma_i}{\sqrt{c_i}}$	$\pi_i\sigma_i\sqrt{c_i}$
Easy	.6	25	5	160,000	3200	400	1.04	48	1200
Difficult	.2	36	6	90,000	1800	300	5.00	10	360
Very difficult	.2	900	30	90,000	1800	300	25.00	2	1800
	1.0								

[a] $C_0 = 25[1800+360+1200]-50[900+36+25] = \$35,950$
$C_1 = 5[360+1200]-50[36+25] = \$4750$
$n_{\text{easy}} = 48[(35,000+3050)/1560]-50 = 1121$
$n_{\text{diff.}} = 10(24.4)-50 = 194$
$n_{\text{very diff.}} = 0$
Check: $(25)(1121)+(36)(194) = \$35,009$

The computations indicate that the optimum sample design is to take 1121 cases from the easy stratum, or 85% of the total sample of 1315; the remaining 15% comes from the difficult stratum. *No* cases are taken from the very difficult stratum. The variance of this optimum design is only one-third that of a proportional sample. (The variance computations are given in Table 6.14 and Example 6.16.)

Although costs and variances differ from study to study, a Bayesian analysis frequently leads to omission of the most difficult high-cost respondents after reasonable efforts have been made to obtain cooperation. To omit a stratum entirely, it is necessary that there be some prior information; if there is none, omitting the stratum will lead to a sample bias. Either explicitly or implicitly, most researchers express some prior beliefs about the omitted respondents or assume that they are similar to those who are interviewed.

Many researchers, having omitted the most difficult stratum, sample the remaining two strata proportionately. Our sample then would consist of 1261 cases with 946, or 75%, from the easy stratum and 315 from the difficult stratum. This would increase the variance by only 3% and make the data processing and analysis easier (see Table 6.14).

6.6 BOTH FIXED AND VARIABLE COSTS DIFFER BY STRATA WITH PRIOR INFORMATION

If both fixed and variable costs differ by stratum, the solution by Ericson (23) again consists of first deciding what strata are included and then determining the optimum sample size for the included strata.

Using the same notation as in Chapter 4, for the ith project let

M = total budget of funding agency
W_i = importance of research to funding agency of ith project
C_i = fixed cost of research of ith project
c_i = per-unit cost of research of ith project
P_i = amount of prior information about the ith project = $1/V_i^2$
$1/\sigma_i^2$ = the value of a unit of information.

Then, if S is the set of included projects, then for $i \epsilon S$,

$$n_i = \frac{W_i \sigma_i}{\sqrt{c_i}} \left[\frac{M - \sum_{j \epsilon S} C_j + \sum_{j \epsilon S} c_j \sigma_j^2 P_j}{\sum_{j \epsilon S} W_j \sigma_j \sqrt{c_j}} \right] - \sigma_i^2 P_i . \tag{6.5}$$

The feasible set of included projects must fulfill the condition,

$$M > C_S \equiv \sum_{j \epsilon S} C_j + \max_{i \epsilon S} \left[\frac{\sqrt{c_i} \sigma_i P_i}{W_i} \right] \sum_{j \epsilon S} W_j \sigma_j \sqrt{c_j} - \sum_{j \epsilon S} c_j \sigma_j^2 P_j \tag{6.6}$$

The maximum value of information under this constraint is

$$V_S(M) = \left[\sum_{i \notin S} \frac{W_i^2}{P_i} + \frac{\left[\sum_{i \epsilon S} W_i \sqrt{c_i} \sigma_i \right]^2}{M - \sum_{i \epsilon S} C_i + \sum_{i \epsilon S} c_i \sigma_i^2 P_i} \right]^{-1} \tag{6.7}$$

For a given M, certain sets are eliminated as being too costly. From all remaining sets, the set S is chosen to maximize V.

After the optimum funding decision is made, there are still two gaps in the information obtained:

1. For projects that are not funded at all, we know nothing that we did not know already. This is expressed by the first term in the denominator of Formula (6.7), $\sum_{i \notin S} (W_i^2/P_i)$.
2. For projects that are funded completely or partially, there will be a residual level of uncertainty expressed by the other term in the denominator of Formula (6.7).

Example 6.12 Funding Agency Granting Decision

Suppose a granting agency with \$100,000 to allocate has four project applications and assigns values W_i to the projects, based on the agency's priorities. The fixed and variable costs, C_i and c_i, are supplied by the grant applicants. The values for P_i and σ_i^2 are obtained from technical experts. Table 6.9 gives the assigned parameter values. Table 6.10 gives the necessary computations and Table 6.11 ranks the sets by increasing C_S.

Table 6.9
Values Assigned to Parameters for Four Project Applications

i	1	2	3	4
W_i	2	3	4	2
C_i	25,000	30,000	40,000	50,000
c_i	9	16	36	16
P_i	8	2	4	4
σ_i^2	100	400	100	100

Table 6.10
Computations for Project Selection

i	1	2	3	4
$W_i\sigma_i\sqrt{c_i}$	60	240	240	80
$\sqrt{c_i}\sigma_i P_i/W_i$	120	53.33	60	80
$c_i\sigma_i^2 P_i$	7200	12,800	14,400	6400
W_i^2/P_i	.5	4.5	4	1

Table 6.11
Sets Ranked by Increasing Fixed Costs

	(1) $\sum_{i\epsilon S} C_j$	(2) $\max_{i\epsilon S} \frac{\sqrt{c_i}\sigma_i P_i}{W_i}$	(3) $\sum_{i\epsilon S} W_i\sigma_i\sqrt{c_i}$	(4) $\sum_{i\epsilon S} c_i\sigma_i^2 P_i$	(5) C_S (1) + (2)(3) − (4)	(6) $V_S(M)$
0(0000)	0	0	0	0	0	.100
1(1000)	25,000	120	60	7,200	25,000	.105
2(0100)	30,000	53.33	240	12,800	30,000	.161
3(0010)	40,000	60	240	14,400	40,000	.148
4(0001)	50,000	80	80	6,400	50,000	.110
5(1100)	55,000	120	300	20,000	71,000	.157
6(0110)	70,000	60	480	27,200	71,600	.181
7(1001)	75,000	120	140	13,600	78,200	.111
8(1010)	65,000	120	300	21,600	79,400	.141
9(0101)	80,000	80	320	19,200	86,400	.141
10(0011)	90,000	80	320	20,800	94,800	.120
11(1101)	105,000	120	380	26,400	124,200	–
12(1110)	95,000	120	540	34,400	125,400	–
13(0111)	120,000	80	560	33,600	131,200	–
14(1011)	115,000	120	380	28,000	132,600	–
15(1111)	145,000	120	620	40,800	178,600	–

Looking at Table 6.11, it is seen immediately that sets 11–15 are eliminated because they are too costly. Computing $V_S(M)$ from Formula (6.7) for the remaining sets, it is seen that set S_6, consisting of the second and third projects, gives the maximum information.

From Formula (6.5), the optimum sample sizes for the selected projects are

$$n_2 = 15\left(\frac{30{,}000 + 27{,}200}{480}\right) - 800 = 988$$

$$n_3 = \frac{20}{3}(119.17) - 400 = 395.$$

It should be apparent that Formulas (6.4) and (6.5) are identical except for the differences in notation. This illustrates that the procedures for determining an optimum sample, once the included strata are determined, depends only on variable, and not on fixed, costs per stratum.

6.7 ESTIMATION PROCEDURES AND SAMPLING ERROR COMPUTATION FOR DISPROPORTIONATE SAMPLES

The optimum sampling procedures described in this chapter make it necessary to weight the data to obtain unbiased estimates of the total population. With current computers, this is an easy task if provisions are made when the study is being planned. Since, for the usual social science study, the data will be split in many different ways, the simplest procedure is to put the weight into each data record.

The weight of each unit in the hth stratum is simply the ratio N_h/n_h, the number of elements in the population divided by the sample size in the hth stratum. Thus, for the total population, a sample total is estimated by summing over all observations:

$$x = \sum_h N_h \bar{x}_h$$

where

$$\bar{x}_h = \frac{\sum^{n_h} x_{ih}}{n_h}. \tag{6.8}$$

That is, the population total is merely the weighted sum of strata means, and the population mean is this total divided by N, the population size.

$$\bar{x} = \frac{x}{N} = \frac{\Sigma_h N_h \bar{x}_h}{N} \tag{6.9}$$

Example 6.13 Estimation from Hospital Sample (Example 6.7 Continued)

Suppose one has selected an optimum sample of hospitals based on the number of employees (see Table 6.4) and now wishes to estimate from sample results the proportion of all hospitals with pension plans for employees. The computations are given in Table 6.12. It may be seen that the weighted estimate of 55% is much smaller than the unweighted estimate of 67%. This is the case because the large hospitals that are more likely to have pension plans are oversampled.

The 67% estimate is not meaningless, however, but may be as important as, or more important than, the weighted estimate. If all employees in a given hospital have the same benefits, the higher 67% is an estimate of the number of hospital employees covered by pension plans. For this estimate, the optimum sample is self-weighting. To generalize, if one is concerned about the institutions themselves, regardless of size, and one has used disproportionate sampling, then weighting is required. If one is not concerned about institutions per se but about the clients or customers they serve, their total revenues or employees, then a sample selected proportionate to some size measure need not be weighted.

It is often the case that both kinds of information are required from the same survey. It is then essential that the researcher describe, for each variable separately, how it was estimated.

It is unlikely that anyone who has carefully chosen an optimum sample design would ignore estimation problems. This is more likely to occur when the stratification has been arbitrary and the total population is difficult to define.

Table 6.12
Estimation of Proportion of Hospitals with Pension Plans (Hypothetical)

Number of beds	N_i	n_i	N_i/n_i	x_i Sample hospitals with pension plans (hypothetical)	$p_i = \frac{x_i}{n_i}$	$\frac{N_i x_i}{n_i}$
Under 50	1614	36	44.8	16	.44	717
50–99	1566	71	22.1	36	.51	794
100–199	1419	120	11.8	66	.55	780
200–299	683	93	7.3	55	.59	404
300–499	679	231	2.9	162	.70	476
500 and over	609	449	1.4	337	.75	457
Totals	6570	1000		672		3628

$$\Sigma \frac{p_i N_i}{n_i} = \frac{3628}{6570} = .55 \qquad \frac{\Sigma x_i}{n} = \frac{672}{1000} = .67$$

Then, there is an unfortunate tendency to dismiss the stratification and treat the sample as self-weighting. The results are then almost impossible to interpret.

Example 6.14 The Sample of Fifteen Cities (Example 1.3 Continued)

In the initial design of the sample of 15 major American cities, for the study of riots, the plan had been to publish a report about each city separately and a summary report of findings that were general in all cities. Because of extreme time pressures, the initial report combined the data for all 15 cities into one sample, without weighting. Since the sample sizes were about the same in all cities, although the cities varied in size by a factor of 20 from the largest to the smallest, the combined results gave greater weight to the smaller cities. The results could be unbiased only if there were no differences by city size. The researchers claimed that they observed no major differences between cities in some limited analyses, but no details were presented.

SAMPLING ERROR COMPUTATION

For a stratified sample, the sampling variance is computed separately for each stratum. If random sampling has been used within stratum, the regular random sampling error formulas may be used, along with a finite correction factor if the sample within the stratum is a large fraction of the total population. A full discussion of sampling error computation is given in Chapter 8. The variance for a *total* summed over all strata is found by multiplying the variances of strata means by the squares of the strata population sizes, and summing over strata.

$$\sigma_x{}^2 = \sum_h N_h{}^2 \, \sigma^2_{h\bar{x}} \qquad (6.10)$$

where

$$\sigma^2_{h\bar{x}} = \frac{(1-f_h)\,\sigma_h{}^2}{n_h}$$

and $\sigma_h{}^2$ is the population variance in the hth stratum.

$$f_h = \frac{n_h}{N_h}$$

$1-f_h$ is the finite correction factor.

The variance for a mean is merely the variance for the total divided by N^2.

$$\sigma_{\bar{x}}{}^2 = \frac{1}{N^2} \sum_h N_h{}^2 \, \sigma^2_{h\bar{x}} = \sum_h \pi_h{}^2 \, \sigma^2_{h\bar{x}} \qquad (6.11)$$

Note that Formulas (6.10) and (6.11) are only for totals and means and cannot be used for more complex statistics, such as ratio estimates and regression coefficients.

Table 6.13
Payroll Variance Computations for Stratified Samples

		Proportional		Optimum payroll		Employees	
Number of beds	$\pi_h^2 \sigma_h^2$	$(1-f_h)$	$\frac{(1-f_h)}{n_h}$	$(1-f_h)$	$\frac{(1-f_h)}{n_h}$	$(1-f_h)$	$\frac{(1-f_h)}{n_h}$
Under 50	2,027	.85	.0035	.98	.0288	.98	.0272
50–99	5,656	.85	.0036	.96	.0168	.95	.0134
100–199	19,170	.85	.0039	.93	.0089	.92	.0077
200–299	19,625	.85	.0082	.84	.0079	.86	.0092
300–499	64,358	.85	.0083	.72	.0037	.66	.0029
500 and over	451,733	.85	.0091	.17	.00034	.26	.00058
$\Sigma \frac{\pi_h^2 \sigma_h^2 (1-f_h)}{n_h}$			4908		871		

Example 6.15 Computing Variances in Table 6.4 (Example 6.8 Continued)

Table 6.13 gives the values for computing the variances in Table 6.4, using Formula (6.11) and data from Tables 6.3 and 6.4. Note that the values of $(N_h^2)/N^2$, or π_h^2, and σ_h^2 are the same for each set of computations but that n_h and f_h change, so that one recomputes $(1-f_h)/n_h$ to compare sample designs.

Example 6.16 Computing the Posterior Variance from an Optimum Bayesian Sample (Example 6.11 Continued)

Exactly the same procedure as in Example 6.14 is used to compute the posterior variance, with one modification: The n_h in each stratum are the equivalent samples and are the sums of the new samples selected from each

Table 6.14
Computation of Posterior Variances with Prior Information

		Equivalent posterior n		
Stratum	$\pi_h^2 \sigma_h^2$	Optimum	Proportional omit very difficult	Proportional
Easy	57,600	1121+50	946+50	104+50
Difficult	3600	194+50	315+50	35+50
Very difficult	3600	0+50	0+50	35+50
Totals		1465	1411	324
$\Sigma \frac{\pi_h^2 \sigma_h^2}{n_h}$		135.9	139.7	458.7
Ratio $\sigma^2/\sigma^2_{opt.}$		1	1.03	3.37

stratum plus the equivalent samples based on prior information. We saw in Example 6.11 that prior information was the equivalent of samples of 50 in each stratum. The computations are given in Table 6.14. Note that the finite correction factor is ignored, since we are sampling from a very large population and the sampling rate is very small.

6.8 SUMMARY

Stratified sampling procedures are used when the strata are of primary interest, or when variances, costs, or prior information differ by strata. The purpose of stratified sampling is to provide the smallest sampling error, hence the most information, for the available resources.

Proportional stratified sampling procedures have little effect if general population samples of individuals are being studied, for means between different strata usually do not differ greatly. Major increases in efficiency are obtained only from disproportionate samples in which the strata are sampled in direct proportion to each stratum size and the square root of variance, and are sampled in inverse proportion to strata per-unit costs. When strata variances are unknown, the use of size measures often yields good approximations of variances for the strata. A simple procedure for substantially reducing the variance in a sample of institutions or firms is to take all the largest ones.

When the strata are of prime interest, the most efficient sample design is to make all stratum sample sizes equal. Frequently, a compromise must be made between the most efficient sample for estimating parameters of the total population and that for differences between strata.

The use of Bayesian analysis for stratified sampling indicates that smaller samples should be taken from strata for which there is available prior information. In the extreme case, if costs are high, no sampling will be done in some strata. This helps explain why researchers are generally willing to settle for less than total cooperation. Bayesian procedures are also useful in explaining the funding decision of granting agencies on the basis of both fixed and variable costs of competing projects as well as the value of information and prior knowledge.

6.9 ADDITIONAL READING

All the classic texts in sampling give excellent discussions of stratification, ignoring prior information. See Cochran (14, Ch. 6, 6a), Hansen, Hurwitz, and Madow (35, Ch. 6), and Kish (40, Ch. 3, 4).

The Bayesian approach is given in two important papers by Ericson (23, 24) in the *Journal of the American Statistical Association.*

7

Multistage Samples

It is unlikely that many readers of this book will ever design a national population sample of the United States or any country, but some might like to know how it is done. There is a possibility that there may arise a need to select a sample of households or individuals from a state or part of a state. Even more likely is the need to select special populations, such as students in universities or employees in specified occupations. The common thread of all these samples is that they require selection in several steps, since complete population lists are not available. In this chapter, we discuss the three major issues and procedures necessary to draw a multistage sample. These issues are:

1. Accounting for size differences of natural clusters
2. Obtaining published data for sample selection
3. Listing procedures where data are unavailable

7.1 AREA SAMPLING

Suppose one needs a state sample of individuals to determine attitudes toward the governor or legislature. There is no central list of all residents age 18 and over in the state, even of registered voters. Information on voters is available at each county, arranged by ward and precinct or however an election district is defined. Interviewing is to be done face-to-face, so clustering is essential for efficiency. Stratification is of less importance but will be considered if it does not significantly add to costs. The researcher knows that a sample of size n is required. How does he begin?

The first step is to outline the selection stages:

1. Primary Sampling Units (PSUs) are selected from the counties or Standard Metropolitan Statistical Areas (SMSAs) in the state.
2. Places are selected within the PSUs. These are cities, towns, unincorporated places, or minor civil divisions.
3. Segments are selected within places. In cities, these segments are generally blocks. In rural places, these are areas, such as Census Enumeration Districts, that have natural or man-made boundaries that are easily identified—rivers, roads, railroad tracks, and so on.
4. Voters are selected within segments.

SELECTION OF PSUs

Since the primary sampling units are listed in atlases and other publications, it is very easy to select a simple random sample or systematic sample. This is not an efficient sampling procedure, however. The SMSAs and counties vary enormously in size, from SMSAs with populations in the millions to the smallest rural counties with populations under 5000. If there is a large variation between PSU (or any cluster) sizes, the result is a very sharp increase in sampling variances if the PSUs or clusters are selected at random. It is worth stopping for a moment to see why this is so.

With a fixed sample that is self-weighting, if some clusters are very large, others must be much smaller. The major increase in sampling error occurs from the clusters having small samples. It is easy to prove that the optimum sample with the smallest error occurs (except for very large clusters) when the sample size is the same in all clusters. Example 7.1 illustrates the problems of sampling PSUs of differing sizes with equal probabilities.

Example 7.1 Selection of PSUs with Equal Probabilities

Suppose one wishes to select a sample of 1000 households from the total of 3.5 million households in Illinois. The overall sampling rate is 1 in 3500. If one first selects with equal probability one-fifth of all the counties in the state, the sampling rate within each county will be 1 in 700, since $(1/5)(1/700) = 1/3500$.

A systematic selection was made of 20 of the 102 counties in the state; the expected sample sizes for each county are given in Table 7.1. Shockingly, instead of a sample of 1000, the expected sample size for the 20 selected counties combined was only 558. This is because half the households in Illinois live in

Table 7.1
Expected Sample Sizes from Selected Counties

County	Number of households	Sample size (HH/700)	$1/n_i$
Boone	7,800	11	.09
Cass	4,800	7	.14
Clinton	8,000	11	.09
DeKalb	19,600	28	.04
Edwards	2,600	4	.25
Fulton	14,400	21	.05
Hancock	7,900	11	.09
Jackson	16,100	23	.04
Johnson	2,600	4	.25
Lake	102,900	147	.01
Logan	10,100	14	.07
Macoupin	14,900	21	.05
Massac	4,900	7	.14
Morgan	11,400	16	.06
Piatt	5,100	7	.14
Randolph	9,600	14	.07
Sangamon	54,400	78	.01
Stephenson	15,900	23	.04
Warren	6,900	10	.10
Will	70,700	101	.01
Totals		558	1.74

Cook County (Chicago), and Cook County was not selected. It is obvious that any sensible sample of the state must include this county. (Detailed discussion of the handling of very large PSUs is given in the next section.)

The revised sample design calls for a sample of about 500 from the remaining 1.75 million households so the sampling rate within each county will still be 1 in 700. Now the sample is a little larger than desired, due to the large variation in sample sizes within each county.

If one assumes that the population variances of the variable being studied are the same for each county, the sampling variance for a sample of 20 counties, each with a sample of 25 households, will be

$$\sum_{1}^{20} \frac{\sigma^2}{25} = \frac{20\,\sigma^2}{25} = .8\,\sigma^2.$$

The variance for the sample in Table 7.1 is found from:

$$\sum \frac{\sigma^2}{n_i} = \sigma^2 \sum_i \frac{1}{n_i}$$

and is seen to be 1.74 σ^2 or *more than twice as large as a sample in which all counties are sampled equally.* It may be seen that large contributions to total variance are attributable to Edwards, Johnson, Cass, Massac, and Piatt Counties, where the sample sizes consist of only four or seven cases.

In addition to large increases in sampling variance due to small sample sizes in a PSU, data-collection procedures also suffer. It is difficult to recruit an interviewer, and generally costs are higher and quality is lower if the interviewer has an assignment that is too small.

For these reasons, it is desirable to try to get approximately equal samples from each PSU, except the very largest ones. But, if one wants a self-weighting sample in which all respondents have the same probability of selection, it is not possible to select PSUs with *equal* probability. Instead, the selection of PSUs must take into account their differing sizes. Such procedures are widely used and are called "sampling with *probabilities proportionate to size,*" or "sampling pps."

7.2 SAMPLING WITH PROBABILITIES PROPORTIONATE TO SIZE

In this section, we describe a procedure for sampling differing sizes of clusters with probabilities proportionate to size, and discuss why sampling pps leads to approximately equal sample sizes in each cluster.

Essentially, the procedure involves assigning to each cluster a sequence of random numbers equal to its size and then sampling systematically. In greater detail, the following steps yield a sample selected pps:

1. The PSUs or clusters are arranged in the desired order to obtain possible benefits from stratification, that is, sorted by region, race, economic, and other variables. (Arranging by size is unnecessary because sampling pps will give the same results.)
2. The size measure for each PSU or cluster is obtained from U.S. census data or other sources.
3. The size measures are cumulatively summed over clusters.
4. The sampling interval is determined as the total cumulative sum of the size measures divided by m, the number of clusters desired.
5. A random start is selected, and the selection numbers are found as r, $r+s$, $r+2s$, $r+3s$, . . . ,$r + (m-1)s$, where r is the random start and s is the sampling interval.
6. A cluster is selected if the selection number falls into its sequence of numbers; that is, the selection number is greater than the cumulative sum

of all previous clusters, but less than or equal to the cumulative sum including the designated cluster.

Example 7.2 Selection of Illinois Counties with Sampling PPS

Continuing Example 7.1, suppose we still want a sample of 20 counties, in addition to Cook County, but we now wish to select them with probabilities proportionate to size. The data are given in Table 7.2. Counties are arranged in ascending order by median income of families to achieve some implicit social class stratification. The cumulative size measure is 1,736,103, the number of households in the state in 1970, excluding Cook County. Dividing this number by 20, the desired number of PSUs gives a sampling interval, s = 86,805. A random start r is selected between 1 and 86,805; the selection is 75,116. The selection numbers then are computed as:

r	75,116	$r+10s$	943,166
$r+s$	161,921	$r+11s$	1,029,971
$r+2s$	248,726	$r+12s$	1,116,776
$r+3s$	335,531	$r+13s$	1,203,581
$r+4s$	422,336	$r+14s$	1,290,386
$r+5s$	509,141	$r+15s$	1,377,191
$r+6s$	595,946	$r+16s$	1,463,996
$r+7s$	682,751	$r+17s$	1,550,801
$r+8s$	769,556	$r+18s$	1,637,606
$r+9s$	856,361	$r+19s$	1,724,411

The indicated counties in Table 7.2 are then selected.

Very Large PSUs. It is likely, in selecting a systematic sample of PSUs, that there will be some larger than the sampling interval. These PSUs must fall into

Table 7.2
Selection of Counties PPS

County	Median income ($)	Number of households	Cumulative	Selected counties
Pulaski	4,933	2,992	2,992	
Pope	5,046	1,296	4,288	
Alexander	5,471	4,348	8,636	
Hardin	5,704	1,762	10,398	
Hamilton	5,873	3,183	13,581	
Brown	6,030	1,947	15,528	
Johnson	6,660	2,558	18,086	
Franklin	6,833	14,356	32,442	
Saline	6,857	9,550	41,992	
Wayne	6,884	6,058	48,050	

(continued on next page)

Table 7.2 (continued)

County	Median income ($)	Number of households	Cumulative	Selected counties
Massac	7,025	4,889	52,939	
Richland	7,082	5,592	58,531	
Edwards	7,087	2,571	61,102	
Calhoun	7,095	1,894	62,996	
Union	7,115	5,442	68,438	
White	7,202	6,197	74,635	
Pike	7,214	6,762	81,397	X75,116
Gallatin	7,288	2,545	83,942	
Jefferson	7,292	10,806	94,748	
Fayette	7,306	6,835	101,583	
Scott	7,357	2,070	103,653	
Greene	7,382	5,783	109,436	
Clay	7,425	5,168	114,604	
Lawrence	7,538	5,993	120,597	
Marion	7,542	13,502	134,099	
Schuyler	7,556	2,840	136,939	
Washington	7,579	4,707	141,646	
Montgomery	7,643	10,362	152,008	
Williamson	7,687	17,252	169,260	X161,921
Cumberland	7,719	3,221	172,481	
Shelby	7,792	7,513	179,994	
Cass	7,795	4,832	184,826	
Effingham	7,852	7,584	192,410	
Bond	7,879	4,581	196,991	
Perry	7,882	6,849	203,840	
Hancock	7,894	7,907	211,747	
Jackson	7,918	16,143	227,890	
Wabash	7,918	4,390	232,280	
Edgar	8,020	7,566	239,846	
Jasper	8,031	3,458	243,304	
Clark	8,046	5,699	249,003	X248,726
Macoupin	8,091	14,920	263,923	
Henderson	8,307	2,800	266,723	
Stark	8,336	2,472	269,195	
Warren	8,369	6,938	276,133	
Jersey	8,403	5,546	281,679	
McDonough	8,467	10,265	291,944	
Mason	8,472	5,637	297,581	
Crawford	8,555	7,124	304,705	
Christian	8,556	12,108	316,813	
Menard	8,563	3,252	320,065	
Clinton	8,577	7,987	328,052	
JoDaviess	8,613	6,724	334,776	
Fulton	8,620	14,403	349,179	X335,531
Mercer	8,683	5,572	354,751	
Iroquois	8,723	10,886	365,637	

Table 7.2 (continued)

County	Median income ($)	Number of households	Cumulative	Selected counties
Randolph	8,818	9,553	375,190	
Adams	8,879	23,105	398,295	
Bureau	8,884	12,525	410,820	
Moultrie	8,987	4,234	415,054	
Coles	8,899	15,240	430,294	X422,336
Ford	9,035	5,402	435,696	
Carroll	9,092	6,360	442,056	
Marshall	9,141	4,358	446,414	
Morgan	9,323	11,368	457,782	
DeWitt	9,330	5,857	463,639	
Logan	9,330	10,099	473,738	
Monroe	9,352	5,757	479,495	
Putnam	9,364	1,637	481,132	
Vermilion	9,449	32,040	513,172	X509,141
Douglas	9,487	6,185	519,357	
Knox	9,506	20,081	539,438	
St. Clair	9,547	86,347	625,785	X595,946
Henry	9,554	17,226	643,011	
Livingston	9,612	12,426	655,437	
Lee	9,636	10,940	666,377	
Piatt	9,818	5,108	671,485	
LaSalle	9,953	35,767	707,252	X682,751
Woodford	9,968	8,337	715,589	
Stephenson	9,987	15,931	731,520	
Whiteside	10,014	19,363	750,883	
Champaign	10,147	47,361	798,244	X769,556
Ogle	10,166	13,454	811,698	
McLean	10,183	31,874	843,572	
Madison	10,249	78,470	922,042	X856,361
Sangamon	10,302	54,374	976,416	X943,166
Macon	10,325	40,808	1,017,224	
Kankakee	10,445	27,942	1,045,166	X1,029,971
Rock Island	10,581	53,587	1,098,753	
Peoria	10,635	63,323	1,162,076	X1,116,776
DeKalb	10,735	19,646	1,181,722	
Tazewell	10,787	36,474	1,218,196	X1,203,581
Grundy	10,982	8,317	1,226,513	
Boone	11,051	7,778	1,234,291	
Winnebago	11,058	76,716	1,311,007	X1,290,386
Will	11,791	70,688	1,381,695	X1,377,191
Kendall	11,929	7,485	1,389,180	
Kane	11,947	74,642	1,463,822	
McHenry	11,965	33,083	1,496,905	X1,463,996
Lake	13,009	102,947	1,599,852	X1,550,801
DuPage	14,458	136,251	1,736,103	X1,637,606
				X1,724,411

the sample with certainty. In the example, in addition to Cook County, Lake and DuPage Counties—which are also part of the Chicago Standard Metropolitan Area—are certainty counties. Due to the random selection, Lake County falls in once and DuPage County falls in twice. This means that a double sample will be selected from DuPage County, so, if approximately 25 respondents are sampled from all other counties, there will be about 50 cases in DuPage. Note that, since there is a double sample in one county, there are only 19, not 20, different counties in the sample.

A slightly better method, after computing the initial sampling interval, is to place all the certainty PSUs into a separate stratum. The sample selected from the certainty PSUs will depend only on the overall sampling rate and will be larger than for the noncertainty PSUs. For the remaining noncertainty PSUs, a new sampling interval is determined as the total of the size measures of the noncertainty PSUs divided by $(m-c)$ where c is the number of certainty PSUs.

Example 7.2 (Continued) Selection of Illinois Counties

The cumulative size measure of noncertainty PSUs is seen to be 1,496,905, excluding Lake and DuPage Counties. Suppose we want 20 *distinct* counties in addition to Cook County; then $(m-c)=20-2=18$. The re-computed sampling interval s is now 1,496,905/18, or 83,161. It is now necessary to check the list again to see if any new counties become certainty counties as a result of the reduced sampling interval. St. Clair County, which was almost a certainty county before, now becomes one, and the sampling interval is again re-computed by subtracting the size measure for St. Clair County:

$$s = 1{,}410{,}558/17 = 82{,}974.$$

The process continues until, as in this example, there are no additional certainty counties. The remaining noncertainty counties are then selected pps, as already described.

7.3 HOW SAMPLING PPS GIVES EQUAL SAMPLE SIZES IN NONCERTAINTY CLUSTERS

Since PSUs are selected with probabilities proportionate to their size and since the sample is self-weighting, the probabilities of selection within each PSU, given that the PSU has been selected, vary inversely with the size of the PSU.

Let

$$P_T = \text{overall probability of selection} = \frac{n}{N}$$

$$P_{\text{PSU}} = \text{probability of a PSU being selected} = \frac{MOS_{\text{PSU}}}{N/m}$$

where MOS_{PSU} is the measure of size for the PSU.

Then, ignoring for the moment the intermediate stages, the probability of selection within the PSU P_W must be such that

$$P_{\text{PSU}} \times P_W = P_T.$$

Solving for P_W,

$$P_W = \frac{P_T}{P_{\text{PSU}}} = \frac{n}{N} \times \frac{N}{mMOS_{\text{PSU}}} = \frac{n}{m} \frac{1}{MOS_{\text{PSU}}}. \tag{7.1}$$

The estimated number of cases from any PSU is the number of units in the PSU times the probability of selection within, or:

$$n_c = \left(MOS_{\text{PSU}}\right)\left(\frac{n}{m} \frac{1}{MOS_{\text{PSU}}}\right) = \frac{n}{m}.$$

Thus, regardless of the initial size of the PSU, the expected number of cases is the same, n/m.

For certainty PSUs, $P_W = P_T$ since $P_{\text{PSU}} = 1$. The expected n_c in any certainty PSU is $(MOS_{\text{PSU}})(n/N)$.

Example 7.2 (Continued) Expected Sample Sizes in Selected Sample of Illinois Counties

We shall continue to assume that a total sample of 1000 households for the state is desired, so the overall sampling rate is about 1 in 3500. As before, half the sample, or 500 cases, will be in Cook County. For the other certainty counties, $P_W = P_T = 500/1{,}736{,}103$, or .000288. The estimated sample size is 39 households in DuPage County, 30 in Lake County, and 25 in St. Clair County, based on this sampling rate.

The remaining sample of 406 households will be in 17 counties, giving each county a sample size of 406/17, or about 23–24 households. To illustrate this, consider the first county selected, Pike County. Using our corrected sampling interval s = 82,974, the probability of Pike County being selected is P_{PSU} = 6762/82,974 = .0815. Since P_T = .000288,

$$P_W = \frac{P_T}{P_{\text{PSU}}} = \frac{.000288}{.0815} = .00353.$$

The expected sample n_c from Pike County is (.00353) (6762) = 23.9.

7.4 SAMPLING PPS AT SEVERAL STAGES

The same pps procedures for selecting PSUs are used also for the selection of places within PSUs and for segments or blocks within places.

Let

P_T = overall probability of selection

P_{PSU} = probability of PSU selection

P_{place} = probability of place selection
P_{block} = probability of block selection
P_{house} = probability of household selection

and let

m_1 = desired number of PSUs
m_2 = desired number of places per PSU
m_3 = desired number of blocks per place
n = desired total sample size
N = total population

Then $n/m_1 m_2 m_3$ will be the expected cluster size/block in a PSU.

$$P_T = \frac{n}{N}$$

$$P_{\text{PSU}} = \frac{MOS_{\text{PSU}}}{N/m_1}$$

$$P_{\text{place}} = \frac{MOS_{\text{place}}}{MOS_{\text{PSU}}/m_2}$$

$$P_{\text{block}} = \frac{MOS_{\text{block}}}{MOS_{\text{place}}/m_3}$$

$$P_{\text{house}} = \frac{P_T}{(P_{\text{PSU}})(P_{\text{place}})(P_{\text{block}})} = \frac{\dfrac{n}{N}}{\dfrac{MOS_{\text{PSU}}}{N/m_1} \cdot \dfrac{MOS_{\text{place}}}{MOS_{\text{PSU}}/m_2} \cdot \dfrac{MOS_{\text{block}}}{MOS_{\text{place}}/m_3}}$$

$$= \frac{n}{m_1 m_2 m_3 \, MOS_{\text{block}}}.$$

So expected cases/block $= MOS_{\text{block}} \, p_{\text{house}} = n/m_1 m_2 m_3$.

At the block level, households and the individuals within households are selected with equal probabilities, using the procedures described in Chapter 3.

Example 7.2 (Continued) Selection of Places and Blocks in Illinois Sample

In a multistage sample, a nested clustering procedure is one where at each stage smaller clusters are selected within the previously selected larger clusters. Assume that, for the general purposes of a study, optimum nested clustering procedures, based on cost and cluster homogeneity (see Chapter 4), indicate three locations within each noncertainty PSU, and two blocks or segments within each place. Since the sample size for noncertainty PSUs is about 24 respondents, there will be about 4 per block.

For certainty PSUs, additional places and blocks are added, but the clustering at the block level remains the same. There is also the possibility of very large places or blocks falling in with certainty, with the result that their sample sizes are larger than expected. Table 7.3 shows the procedures for selecting the places in Clark, Peoria, and DuPage Counties. Blocks and segments are selected in the same way if data are available; if not, there are alternative procedures, discussed in the last section of this chapter.

Table 7.3
Selection of Places for Clark, Peoria, and DuPage Counties

Place	Measure of size (population)	Cumulative MOS	Selected
A. Clark County	s = 16,894/3 = 5,631	r = 1,156	
Anderson township	323	323	
Auburn township	271	594	
Casey city	2,994	3,588	X1,156
Casey township	960	4,548	
Darwin township	379	4,927	
Dolson township	390	5,317	
Douglas township	183	5,500	
Johnson township	387	5,887	
Marshall city	3,468	9,355	X6,787
Marshall township	828	10,183	
Martinsville city	1,374	11,557	
Martinsville township	502	12,059	
Melrose	391	12,450	X12,418
Orange	352	12,802	
Parker	235	13,037	
Wabash	1,608	14,645	
Westfield village	827	15,472	
Westfield township	678	16,150	
York township	744	16,894	
B. Peoria County	s = 195,318/3 = 65,106	r = 1,938	
Akron township	547	547	
Princeville village	1,455	2,002	X1,938
Brimfield village	729	2,731	
Brinfield township	503	3,234	
Chillicothe city	6,052	9,286	
Chillicothe township	180	9,466	
Rome (u)	1,919	11,385	
Elmwood city	2,014	13,399	
Elmwood township	540	13,939	
Hallock township	1,054	14,993	
Hollis township	1,169	16,162	
Bartonville village	7,221	23,383	
Mapleton village	281	23,664	

(continued on next page)

Table 7.3 (continued)

Place	Measure of size (population)	Cumulative MOS	Selected
Pekin city (part)	4	23,668	
Jubilee township	593	24,261	
Kickapoo township	2,136	26,397	
Limestone township	10,654	37,051	
Bellvue village	1,189	38,240	
Norwood village	632	38,872	
West Peoria (u)	6,873	45,745	
Logan township	1,353	47,098	
Hanna City village	1,282	48,380	
Medina township	4,388	52,768	
Millbrook township	636	53,404	
Peoria city	126,963	180,367	X67,044 X132,150
Peoria township	400	180,767	
Princeville township	567	181,334	
Radnor township	945	182,279	
Dunlap village	656	182,935	
Richwoods township	29	182,964	
Peoria Heights village	7,943	190,907	
Rosefield township	933	191,840	
Timber township	1,063	192,903	
Glasford village	1,066	193,969	
Kingston Mines village	380	194,349	
Trivoli township	969	195,318	
C. DuPage County (households)			
	$s = 136{,}251/5 = 27{,}250$	$r = 3{,}338$	
Census Tract 7701	639	639	
8401	5,172	5,811	X3,338
8402	1,728	7,539	
8403	3,351	10,890	
8404	238	11,128	
8405	2,305	13,433	
8406	2,186	15,619	
8407	1,804	17,423	
8408	2,267	19,690	
8409	3,395	23,085	
8410	1,702	24,787	
8411	1,951	26,738	
8412	1,924	28,662	
8413	1,407	30,069	
8414	1,706	31,775	X30,588
8415	2,703	34,478	
8416	1,844	36,322	
8417	1,679	38,001	
8418	2,250	40,251	
8419	1,357	41,608	

Table 7.3 (continued)

Place		Measure of size (population)	Cumulative MOS	Selected
Census Tract	8420	1,224	42,832	
	8421	1,863	44,695	
	8422	1,485	46,180	
	8423	1,091	47,271	
	8424	1,976	49,247	
	8425	1,224	50,471	
	8426	2,719	53,190	
	8427	3,547	56,737	
	8428	2,095	58,832	X57,838
	8429	1,662	60,494	
	8430	1,463	61,957	
	8431	1,085	63,042	
	8432	2,074	65,116	
	8433	1,693	66,809	
	8434	1,200	68,009	
	8435	1,604	69,613	
	8436	2,275	71,888	
	8437	1,816	73,704	
	8438	1,117	74,821	
	8439	1,899	76,720	
	8440	2,924	79,644	
	8441	1,438	81,082	
	8442	3,099	84,181	
	8443	3,243	87,424	X85,088
	8444	1,088	88,512	
	8445	1,407	89,919	
	8446	1,731	91,650	
	8447	1,978	93,628	
	8448	2,958	96,586	
	8449	2,710	99,296	
	8450	2,036	101,332	
	8451	2,628	103,960	
	8452	1,031	104,991	
	8453	1,528	106,519	
	8454	1,272	107,791	
	8455	2,862	110,653	
	8456	2,610	113,263	X112,338
	8457	2,070	115,333	
	8458	3,490	118,823	
	8459	651	119,474	
	8460	2,365	121,839	
	8461	2,260	124,099	
	8462	2,784	126,883	
	8463	5,830	132,713	
	8464	1,351	134,064	
	8465	2,187	136,251	

There are several features of Table 7.3 that deserve attention:

1. The selection of places for Clark and Peoria Counties was on the basis of population instead of households in order to illustrate the use of different size estimates. The census tract data on households were used in DuPage County. Since the correlation between the population and the number of households in a county is very high (about .97), there is no reason for not mixing household and population estimates of size at various stages of selection, as long as the final selection probability is computed to make the sample self-weighting. This might be necessary when multiple sources are used together for size estimates.

2. In the sample of Peoria County, the city of Peoria has about two-thirds of the population of th entire county, and falls in twice. This means that four blocks are sampled in this city. Here, there is no concern about getting three distinct places, as there was with PSUs.

3. For all Standard Metropolitan Statistical Areas, the entire county is divided into census tracts, each containing about 2000 households on the average. Where available, these tracts are very useful for sampling because they are chosen so as to be relatively homogeneous with respect to demographic variables. Some census tracts cross political lines, such as city limits or election districts, but even here, data are available on the number and characteristics of respondents within a census tract and also within the corporate limits of a city. For special studies, other boundaries may be more useful—for example, ward and precinct boundaries for election studies, or boundaries of telephone companies and exchanges when telephone sampling is used.

4. Note that there were five, not three, census tracts selected in DuPage County since, there, the ratio of households to the sampling interval was 136,000/83,000, or about 5/3. In general, for certainty PSUs, the number of places found is $(MOS_{\text{PSU}})(m_2)/s$ where m_2 is the number of places in noncertainty PSUs. The number of blocks per place, and households or individuals per block, do not change.

The extensive discussion of an area population sample presents many of the problems faced in multistage sampling, but there are other populations for which pps procedures are used. The next two examples illustrate such populations.

Example 7.3 Samples of Graduate Students in Five Engineering Fields

In 1963, the National Opinion Research Center of the University of Chicago conducted a study, for the National Science Foundation, of the stipends being received by graduate engineering students in five fields. The sample size was 1250 students in each field. Although the survey was done by mail, schools had to be selected first so that lists of students could be obtained for sampling. It would have been too expensive to obtain lists from all 129 universities offering graduate training in engineering, so a subsample of one-third of the schools, or 43, was decided upon, based on research funds available.

Schools were sampled with probabilities proportionate to the total number of graduate students enrolled in civil, chemical, electrical, and mechanical engineering. The source of the data was *Engineering Enrollments and Degrees, 1960.* This size measure is not exact for any single field but is the best average measure for *all* fields. The data were a few years old, but there were no more recent figures and it was generally believed that there had been no major changes in enrollments during the period 1960–1963.

Since there were about 25,000 graduate students in the four graduate engineering fields, the sampling interval s was 25,000/43, or 581. Of the schools, 14 with enrollments of more than 581 in the four graduate engineering fields fell into the sample with certainty. These schools had a combined enrollment of 10,500 students. The new sampling interval s' was then computed as 14,500/29, or 500. The probability of selection for the noncertainty schools was number of students/500. The overall probability of selection P_T for each field was:

chemical engineering	1 in 2.43
civil engineering	1 in 3.14
electrical engineering	1 in 9.50
mechanical engineering	1 in 4.86
other engineering	1 in 8.08

The sampling rate within a school for a specific field, say chemical engineering, was:

$$\frac{P_T}{P_{\text{school}}} = \frac{500}{2.43 \text{ number of students}}$$

For the noncertainty schools, this was an average of about 25 students per field, or 125 per school, although the actual sample varied from this. The sampling rate for each field in the certainty schools was merely the overall sampling rate P_T.

Example 7.4 Sample of Hospital Service Employees

Suppose one wants a sample of 1500 hospital service employees. Interviewing will be face-to-face, and estimates of costs and cluster effects indicate that a sample of 100 hospitals with about 15 employees per hospital will be optimum. Although the total number of service employees and the number for each hospital is not available currently, there is a high correlation between total employees and service employees. Thus, hospitals will be selected with probabilities proportionate to the total number of employees.

Once the hospitals are selected, they will be contacted to determine the number of service employees at each. From this information, an estimate of the total number of service employees can be made by taking the ratio of service employees to total employees at the 100 hospitals and applying this to the known total number of employees in the population (see Table 6.3). Then, P_T = 1500/total service employees. Once P_T is known, the sampling fraction for each hospital can be determined. One might think that, since none of the hospitals was selected with certainty, the easiest procedure would be to select exactly 15 service employees from each hospital. The reason this is not a good procedure is discussed in Section 7.7, where this example is continued.

7.5 THE DIFFERENCE BETWEEN SAMPLING CLUSTERS PPS AND STRATIFYING BY SIZE

Researchers who have been exposed both to the idea of sampling institutions stratified by size and to treating these institutions as clusters and sampling pps are frequently confused as to the relation between the two procedures. They are *not* the same and are not intended to be used for the same purpose!

The key decision the researcher must make at the beginning of the study is what universe is being analyzed. If the study is one dealing with institutions, such as schools, factories, or hospitals, then the institution is the element and no clustering is involved within it, although there may be geographic clustering of institutions. The information is usually obtained from one or a few informants and is not about these individuals but about the institution. Data may be obtained about students and patients, teachers or other employees, but it will be aggregated rather than treated by individual.

If, on the other hand, individuals within institutions are the members of the population, as in the last two examples, the institutions become clusters for sampling. Similarly, households are sometimes the units of analysis and at other times the final cluster for sampling individuals.

Once it is clear what the universe elements are, the differences between sampling pps and stratifying by size are of less importance. Stratifying by size is the same as sampling pps when every element in the stratum is given the average size. This is a little less precise, particularly for the largest stratum in which there is considerable size variation. Thus, in sampling differing sizes of clusters with the aim of obtaining equal sample sizes, sampling pps is better.

When institutions are the units of the population, the two methods are almost identical. Weighted estimates for a total or stratum are possible using either method; for data presented by separate strata only, no weighting by size stratification is necessary. Stratification by size is used when large institutions are identified but no exact size measures are available.

7.6 WHEN NOT TO SAMPLE PPS

The extensive discussion of sampling pps may lead the reader to believe that this procedure is always followed when clusters differ in size. There are some exceptions. One major exception is when the entire cluster is selected. If the entire cluster is selected, the probability of selection within the cluster is certainty; so, to make the sample self-weighting, clusters also must be chosen with equal probabilities.

Although it would usually be wasteful to sample the entire cluster, it may be necessary to do so for political reasons or it may be no more, or even less, expensive to obtain information from the entire cluster.

Example 7.5 Study of Drug Use among College Students

A sample of 50 classes were selected with equal probabilities at a university, and all students in the class filled out a self-administered form that asked about their use of alcohol and marijuana and other drugs. The costs involved an interviewer's going to the class, passing out forms, and collecting them when finished. Except for additional costs of data processing, there were no additional data-collection costs in surveying the entire class. In fact, subsampling would have been more expensive because it would have involved special procedures for the many different types of classroom arrangements. The anonymity of respondents was also better preserved by having everyone in the room fill out a form rather than having only a few who conceivably might have been identified by some of their answers. Note that this procedure assumes that all students take the same number of classes. This assumption is satisfactory for full-time students, but leads to an undersample of part-time students.

Example 7.6 Study of Attitudes of Unitarians

A sample of Unitarian churches were selected with probabilities proportionate to size; for the selected churches, however, all members were sent questionnaires. The cost of mailing the questionnaires was borne by the individual churches, who felt that all members should have the opportunity to respond. Otherwise, those omitted might have felt slighted. Subsampling also would have been a difficult procedure to explain to the person responsible for the mailing.

Once the returns were in, respondents were subsampled to yield a self-weighting sample. This was done because processing costs per questionnaire were high, due to the large number of open-ended questions. Here, the processing costs were higher than data-collection costs and determined the sample design.

There is another situation in which clusters are sampled with equal probability. If size measures are not available, but the researcher judges that there is very little variation in cluster size, clusters may be selected with equal probability. It is always possible to obtain measures of size by sending someone into an area, but the cost of this procedure may be greater than justified by the small reduction in sampling variance.

Small towns with no apartment buildings are not likely to have much variation in block or segment sizes, nor are rural areas with large and approximately equal-sized farms, such as sections of townships found in the Midwest. On the other hand, rapidly growing suburban areas and rural areas with varying uses may have substantial differences between cluster sizes. Here, the cost of obtaining new size measures through field counting procedures is more than justified by the reduction in variance.

7.7 WHY SAMPLING PPS DOES NOT MEAN EXACTLY EQUAL SAMPLE CLUSTERS

In the discussion and the examples about sampling pps, we have attempted to stress the point that the procedure yields only approximately equal clusters. Intuitively, many researchers would like to use pps procedures until the final stage, then select exactly equal clusters so as to minimize sampling error and eliminate all variance in the workload. Unfortunately, this yields a biased sample if size measures are imperfect. It is seldom that perfect size measures are available. The most common causes of errors in size measures are

1. Changes over Time. Census information is available only periodically and, at the earliest, a year or more passes before information from any census is published. During the interval between the times the data are collected and are used, a cluster may grow or shrink in size.

At one extreme, a block or segment that earlier had housing units might be totally vacant later because the area was transformed for use as a park, an industrial site, or a site for future new housing. In this case, there will be no sampling of the vacant site. As discussed earlier, in Chapter 3, it is a mistake to substitute an adjacent block for the one that now has no households.

A more troublesome problem is to include in a current sample those blocks or segments that had no households when the census data were collected. If new housing is constructed in these blocks or segments, the households living in these units will have no chance of falling into the sample. There are two procedures used to provide for these changes:

1. At the time a sample is being selected, additions and corrections are made to the earlier census data by use of records like building permits, utility installations, and newspaper ads or articles about new developments or other real estate documents. Alternatively, real estate experts may be consulted, or someone may be sent to cruise through undeveloped areas in search of new construction. These methods are appropriate when the sampling is done in one or only a few places, but probably are too costly to use for a national sample or where many PSUs are involved.
2. Another procedure is to attach in advance all blocks or segments with zero households to an adjacent block with some households before sampling blocks. These zero blocks are not listed in census-block statistics tabula-

tions, but are shown on maps. These combinations of blocks are then treated as a single cluster and sampled pps. If this cluster falls into the sample, the interviewer lists housing units on both blocks. Thus, if new housing is built after the census, it will be listed along with housing on the adjacent block. (Of course, most of the time, the zero blocks will continue to have no residential housing.)

This procedure works well when there are many PSUs and when the new housing is relatively scattered, but it is inefficient if a huge new apartment housing development is built on an empty piece of land. Then, if one uses the sampling rate based on the census data, the cluster selected could be many times larger than the average cluster. To avoid these unpleasant surprises, a combination of Methods 1 and 2 is used. At the least, an effort is made to correct older census data for major new housing developments in each PSU, while more gradual changes are handled by combining zero blocks with other blocks.

2. *Available Size Measure May Differ from Required Size Measure.* As in Examples 7.3 and 7.4, the size measure used is highly but not perfectly correlated with the necessary, but unknown, size measure. Another example of this was discussed in Chapter 1, where is was suggested that a national probability sample can be used for multiple purposes. Thus, if PSUs selected for a national sample of households were used also for a national sample of hospital employees, there would be a high but not perfect correlation between households and hospital employees in a county.

Example 7.7 Sample of Hospital Service Employees (Example 7.4 Continued)

Suppose that, in the sample of 100 hospitals selected in Example 7.4, the overall responses indicate that about 43% of all employees are in the service category. Then, to obtain an average of 15 service employees per hospital, the sampling rate would be 15/.43 total employees or 35/total employees. The computations are given in Table 7.4. It may be seen that the actual sample

Table 7.4
Sample of Hospital Service Employees (Hypothetical)

Hospital	Total employees	Total service employees	$\frac{35}{\text{Total employees}}$	Sample *n* service employees
A	1400	650	.025	16
B	850	340	.041	14
C	525	225	.067	15
D	175	90	.200	18
E	90	30	.389	12

selected from a hospital varies around 15, depending on whether the hospital has a higher or lower percentage of service employees; but, as is generally true, the variations are not very large and do not substantially increase the sampling variability.

3. Problems in Data Collection or Processing. Some small fraction of size measures will be incorrect due to human error even if careful quality control procedures have been used. Finally, during data collection, cooperation rates may vary between clusters.

Sampling pps procedures are *self-correcting* for all errors in estimates of size measures if sampling intervals are used throughout, rather than the specification of an exact number of cases. Thus, if a block has doubled its population since the last census, there will be twice as many cases as expected from that block, but all persons on that block (and on all other blocks that have grown) will have the proper probability of selection. Similarly, if a block's population has shrunk because of demolition or vacancies, there will be fewer cases. The same is true of locations and counties and of institutional clusters.

Errors in size measures cause some increase in sampling error, but usually this increase is slight. If one is concerned that the size measures used are very much outdated or inaccurate, a screening survey to get new and better size measures may be justified. In rapidly growing suburban areas, the use of phone directories, as suggested in Chapter 3, should be considered to improve census data.

7.8 OBTAINING PUBLISHED DATA FOR SAMPLE SELECTION

The frequent references already made to census data indicate that, for many multistage sampling procedures, the principal sources of sampling information are the Censuses of Population and Housing, as well as the Censuses of Agriculture, Business, Governments, and Manufactures. In this section are given references based on the most recent censuses. In addition, many other federal and state sources are useful. Although references are to published data, the census data are also widely found in computer tape libraries–it may be more convenient to use them in that form.

POPULATION AND HOUSEHOLD COUNTS FOR COUNTIES, SMSAs, AND LOCALITIES

Population and housing counts are found in many sources. The most readily available are probably the U.S. Census PC (1)-B series, *General Population Characteristics* for each state. For counties, SMSAs, places of 1000 or more population, and county subdivisions–such as townships–data are available for both population and households. The same information is available for the total rural area of each county.

Census Tract and Block Counts. Population and household counts were available for 241 SMSAs in 1970. The data are published in the PHC(1) series, *Census Tracts,* with a separate volume for each SMSA. Block statistics were available for the urbanized portions of these SMSAs and for other selected places. The 278 volumes are published in the HC(3) series, *Block Statistics.*

Population Counts for Places under 1000. Additional information on population, but not households, is available for all incorporated places arranged by county subdivisions in the series PC(1)-A, *Number of Inhabitants.* By subtraction, it is possible to arrive at the population in a county subdivision outside of all incorporated places.

Block Groups and Enumeration Districts. For places for which census tract counts or block statistics are not available, it is possible to obtain data divided into block groups or enumeration districts. An enumeration district is the area covered by one census enumerator (about 250 households) and is smaller than a census tract in population although somewhat larger in geographic area. A block group used in census-by-mail areas is a group of continguous blocks having a population of about 1000. Block group and enumeration district information is not published routinely but is available on first-count summary tapes, for a nominal charge, from the Census Bureau and other sources.

It is probably worth mentioning that, for sampling, it is necessary to have not only the population counts but maps of places, census tracts, blocks, and enumeration districts. These maps are usually attached to published data, but if other sources are used, maps must be obtained from the Census Bureau.

OTHER CENSUSES

The criteria for including the following censuses is that they all contain information about the number of institutions and some measure of size at the county level:

Census of Agriculture. Volume 1, Area Reports–Information by type and size of farm, by county.

Census of Business, Retail Trade. BC-RA Series. Data, by kind of business, on number of establishments, sales, payroll, and employment for SMSAs, counties, and cities for each state.

Wholesale Trade. BC-WA Series. Same data for wholesale establishments.

Selected Services. BC-SA Series. Same data for selected services, such as personal, business and repair services trades, selected amusement and recreation service trades, and hotels, motels, and tourist courts.

Census of Manufactures. Volume 3, Area Statistics. Same data as in Census of Business for industry groups and important individual industries.

Census of Governments. Volume 7, State Reports. Data on governmental structure, employment, and finances, by counties, SMSAs, and major municipalities.

7.9 THE USE OF LISTS IN MULTISTAGE SAMPLING

The use of lists with simple and pseudo-simple random sampling was discussed in Chapter 3. Lists are also used widely in multistage samples with some added steps. If the data collection is to be done by mail, and no additional stratification is required, a list of institutions may simply be sampled pps as in Examples 7.3 and 7.4. If interviewing will be face-to-face, the list must be rearranged into geographic sequence if it is not so already. If the PSUs to be used are those from an existing national probability sample, all other PSUs are discarded. Within the selected PSUs, institutions are sampled pps from the list.

Perhaps the greatest confusion about the use of lists is in sampling individuals and households when both census tract and block information and city directory lists are available. Intuitively, it would seem that a careful design would be to use the tract and block information and then to select households on the block from the city directory. Unfortunately, city directories are not arranged by tract or block but by street. Since the street numbers are arranged in ascending order, they shift from one side of the street to the other, and thus from one block to another.

The easiest thing to do when a city directory is available is to *ignore* census tract and block data, and sample directly from the directory unless some stratification is required. In the directory, the segment consists of a specified number of lines of listing. Again, the number of eligible respondents will vary between clusters, since some lines will be blank or contain nonhousehold listings.

For some studies, a multistage area sample is used, but the sample consists entirely of respondents in special areas, such as lower-income neighborhoods, identified by census tract boundaries or by other means. In this case, it is still possible to use directories, but it is necessary to check each selected directory cluster on the map to see if it falls into the proper area. Although this may be tedious, it is still easier than obtaining new listings.

7.10 OBTAINING NEW LISTS

If one has no list of individuals such as found in a city directory or from institutional files, it is necessary at the final stage to prepare new lists. The discussion here deals with area sampling because this is the most common situation. Lists are usually required for the very largest cities and for rural areas, since directories are available for most medium-sized places. The listing is generally done by the same group who will later do the interviewing. As a rule of thumb for cost estimation, it generally takes about one-half of a day to list any block or segment, including travel time. The listing instructions given here are excerpted from *How to List for an Area Sample* (54) prepared by the Field Department of the National Opinion Research Center, University of Chicago. These instructions cover the major problems faced in listing.

HOW TO LIST

1. A. **Locating Segments in CITIES and TOWNS.** In general, finding the segments in cities is not difficult because the segments usually have readily identifiable boundaries. A city street map will help you to determine the general location of the segment and the best route to get to it. When you reach it, use the map to verify the exact boundaries. These boundaries may be streets, alleys, streams, railroads, city limits or occasionally even property lines. Sometimes street names or one or more of the segment boundaries may not exist or may not have been known to the person who made the segment sketch. In this case, the general shape of the segment and those boundaries which *are* given will guide you in determining whether or not you are in the right block. In all cases *take an exploratory walk around the block before you start any work on the listings* and make certain you have the correct block. *All* the boundaries must agree. For example, if block is rectangular, the names of three boundaries could agree and you could be wrong–i.e., you could be in an adjoining block.

   B. **Locating RURAL Segments.** The boundaries of segments located outside of cities are usually roads, streams, streets or railroads. Sometimes, however, the boundaries may be imaginary lines, such as township lines or county boundaries. In a few cases, the segment boundary may even be an imaginary extension of a road or simply an imaginary line between two identifiable landmarks. The location and determination of such boundaries may call for some ingenuity. We suggest you proceed as follows:

   1. On a highway map note the general location of the segment and determine the best route to travel to reach it. Note the names of roads, rivers, etc., or the presence of such cultural features as schools, churches, etc., as these may prove useful in identifying the segment.

2. Locate on the map the road being used to reach the segment. On reaching what appears to be the segment boundary, verify this by checking the location of actual terrain features and landmarks against their location on the map. Don't depend on one single feature.
3. Determine the extent of the segment from the map before doing any listing. Those portions of the segment boundary that are coincident with identifiable landmarks (such as roads, rivers, etc.) can be identified merely by locating these landmarks. Segment boundaries that are coincident with county or township lines are so noted on the highway map.
4. While there are cases in which boundaries shown on the map no longer exist or have changed location (e.g., a road has been relocated), don't draw such conclusions too hastily. Make sure that you have gone far enough to locate some landmark *on either side of the presumed boundary.*

It is usually possible to locate unnamed roads or imaginary lines such as Minor Civil Division or county lines, section lines, etc., by one or both of the following procedures:

1. Inquire among people living in the vicinity. In most cases, these people will know where the county line or section line runs. Sometimes you may be able to obtain a more detailed map of the area from a local drug store or stationery shop. You can get good map information from the county recorder of deeds and at the office of a local abstractor or title insurance company. Besides local residents, you may find rural mail carriers, delivery men, surveyors, county engineers, and other local authorities helpful.
2. By noting on the map the approximate distance and direction of several points along the segment boundary from identifiable landmarks, using your automobile speedometer and referring to the scale of miles on the map, it will usually be possible to locate such landmarks and from these to establish just where the segment boundary is.

One caution in using county highway maps should be noted. They are not always completely accurate and up-to-date. This is sometimes true of the road indications, more frequently of the cultural detail shown, such as the location of houses. Where you find discrepancies, correct them by drawing in the missing roads, crossing out non-existent roads, showing the location of new construction, etc.

As in the case of the city segments, if you have any doubt at all of the exact boundaries of the segment, you should make an exploratory trip around the segment before listing.

2. Rules to Follow for Accurate Listing

RULE I. *List All Dwelling Units.* Be sure you account for all units in every residential structure.

a. *The definition of a dwelling unit* (DU) follows that of the U.S. Census Bureau.

A room or group of rooms is regarded as a dwelling unit when it is occupied or intended for occupancy as separate living quarters. Living quarters are separate when (1) the actual or intended occupants do not live and eat with any other persons in the structure, *and* when there is *either* . . .

(2) direct access from the outside or through a common hall . . . *or* (3) a kitchen or cooking equipment for the exclusive use of the occupants.

We present this definition of a dwelling unit for two reasons:
First, it will help you find all the DU's in your segment so that you don't miss any.
Second, it will keep you from listing group quarters as dwelling unit.

b. *A partial list of types of dwelling units* (but remember that this list is merely illustrative, and may have missed some places that should qualify as dwelling units):

–*A single house* which is intended for occupancy by only one family.

–*A flat or apartment* in a structure which includes other flats *or* apartments.

–*A basement or attic apartment.*

–*Vacant houses or apartments which could be occupied.*

–*Hotel or motel rooms* which are (1) occupied by "permanent" guests, or (2) occupied by employees who have no "permanent" residence elsewhere. In listing, indicate "permanent guest" or "employee" to help all interviewers who will work in the segment.

–*Residential units under construction.* In listing, indicate that such a unit is under construction.

–Rooms *within* group quarters or an institution (such as a fraternity house or dormitory) *which serve as the permanent* residence of a staff member and satisfy the requirements of the dwelling unit definition.

–*A room in a non-residential structure where there are no other* rooms occupied or intended for residential occupancy. Thus, if there is one room in a warehouse which the caretaker uses for his living quarters, such a room qualifies as a dwelling unit.

–*A trailer, which is used as the permanent residence* of the occupants and not just their vacation residence.

–*A trailer location* in a trailer lot or park in which numbered or otherwise specified spaces are rented. In such a trailer park, list each separate space allocated for one trailer as a DU even if no trailer currently occupies the space–that is, an empty trailer space in a regular trailer park should be treated like a vacant apartment or house.

–*Work camps occupied by seasonal workers.* Even if a worker occupies a unit for less than six months of the year, that unit should be listed, for it may become a permanent residence in the future.

–*Seasonal dwellings.* List summer homes, resort cottages, or other part-time homes which could serve as permanent residences, *even* if you know or suspect that the present occupant's primary residence is elsewhere.

–*Rooms occupied by lodgers* are separate DU's if they meet the definition of a DU presented earlier. You would list the quarters occupied by each lodger.

Thus if you ran across a rooming house in which all the residents lived and ate separately from each other *and* their rooms opened off a common corridor, *you have found a number* of dwelling units, each of which should be listed on a *separate* line of the DU listing sheet.

You have found *only one* dwelling unit, however, if you run into a boarding house at which the residents actually live and eat together. It is a Dwelling Unit only if there are four or fewer boarders living there. If there five or more such residents, you may have run into group quarters, *not* a dwelling unit. Group quarters are described in the next sub-section.

c. For *group quarters* which you *do not* list we follow again the Census Bureau definition. Five or more persons unrelated to the person in charge, automatically transform what might have seemed to be one Dwelling Unit into Group Quarters. The following are group quarters, and are not to be listed:

–*Boarding houses of 5 or more* whose quarters do not qualify as separate dwelling units.
–*College dormitories* (but watch for the dwelling unit of a "house head" or permanent "dormitory counselor" or some such person).
–*Fraternity houses.*
–*Barracks.*
–*Staff quarters in hospitals.*
–*Convents and monasteries.*

d. List of *"non-residential"* units which are *not* to be listed: (However you cannot completely ignore such institutions, since some of them may contain the dwelling unit of a manager, janitor, proprietor, etc.)
–*Hospitals.*
–*Transient hotels or motels.*
–*Penal institutions.*
–*Homes for the aged.*
–*Other institutions which provide care* for residents or inmates.
–*Unoccupied buildings* which have been condemned or which are being demolished.
–*Places of business* such as stores, factories, etc.–but be sure to look for hard-to-find living quarters behind or above or inside business places.
–*Resort cabins* which can never be used for more than a short period each year.

RULE II. *List by Inspection (chiefly).*

Single-family houses can usually be spotted just be looking at the outside of the house. Sometimes, however, what appears to be a one-family house has been converted into a two or three-family house. Since it is not always possible to tell this fact by looking at the exterior of the house you should inspect, briefly, the outside of each apparent one-family house for multiple dwelling unit "signs" (more than one mailbox, for example). You may find it advisable at times to ask people whom you happen to see on the block whether any of the houses (other than the more obvious ones) have more than one family living in them. Do not spend a great deal of time doing this kind of checking unless there is some indication which warrants it.

Most two-family houses are easy to detect; especially the conventional side-by-side type or the structure containing two floors with identical apartments on each.

Structures with multiple dwelling units, especially apartment houses, also are usually easily detectable. To list DU's in an apartment house, walk into the building and check the bells or mailboxes. If there are no bells or mailboxes, walk around the various floors to make sure you include all the units within the building. Sometimes, the information can be obtained by asking the superintendent, janitor or a tenant.

RULE III. *List in Systematic Order.*

We have provided a system which all listers should follow. This is to insure thoroughness in the listing process, and to help in identifying the dwelling units when we return to these segments at a future time.

You may select any *corner* in the segment most convenient to you to begin listing. Do *not* begin listing in the middle of a block. Designate this corner or starting point by an "X" in the sketch of your segment then draw an arrow pointing in the direction in which you walk from this point as you list.

Whenever there is more than one dwelling unit at a given street address, list each dwelling unit in the following order:

a. By apartment number or flat number if such numbers appear on bells, mailboxes, or on doors of individual apartments. Record the lowest number and lowest letter of alphabet first: (e.g., a,b,c, etc., or 1,2,3, etc., or 1-a, 1-b, 1-c, 2-a, 2-b, 2-c, etc.).

b. When apartment numbers are not available, list DU's according to a systematic ordering of mailboxes or doorbells. For instance if you locate six mailboxes in a row, but there are no apparent *numbers,* you may list them as follows:

908 Spring St., left-hand mailbox,
" " " , second mailbox from left
" " " , third " " "

c. When there is no numerical or alphabetical identification of individual rooms, apartments, or flats, you should list in the following order:

(1) Bottom to top, e.g., Basement to ground floor to second.
(2) Right to left within each floor.
(3) Front to the back on any given side of the floor.

This three-dimensional instruction *bottom to top, right to left, front to back,* should permit you to handle all units systematically.

To avoid confusion, do not use the term first floor, because some people think of the first floor as being one flight up and other think of it as the equivalent of the ground floor. Call the street floor the "ground floor;" always use "second floor" for dwellings one flight up.

The terms "right" and "left" should always refer to *your* right and left as you stand in front of the main entrance to the structure and facing the structure. Whenever possible, "north side," "south side," "east side," "west side" are preferable to "right" and "left."

RULE IV. *Cover the Entire Area.*

Every road, street, alley, court, or passageway in your segment should be covered, because there may be dwelling units hidden in them that you might otherwise miss. Since it is more likely to overlook such units, extreme care is all the more essential. Characteristics of residents of remote units may well differ from those of residents of the more conventional or easily accessible units. Enter every alley, street, or road within your segment as you come to it, going down one side and back up the other side.

Dwelling units may be situated in such other unusual places as: in or behind stores, behind other units that face the street, behind the business front of a store building, over garages or stores, in factory yards, etc. It is important to include the dwelling units found in all such places.

In listing for the Census it was found that errors were most often made in the congested areas of large cities and in remote, sparsely settled areas.

In rural segments, begin at one end of each road in the segment as you come to it, listing first all the units on one side, then when you reach the end of the road, reverse your direction and list all dwelling units on the other side, IF other side of the road is within the segment and not across the road from the segment. Include on the list as you come to them (a) every dwelling unit on a road in the segment having its main entry from the road being listed and (b) any other dwelling units in the segment which are not served by a road in the segment, but are more accessible from the road being listed than from any other road in the segment. In this way, every dwelling unit in the segment will be listed once and only once.

In some instances a segment boundary may cut across a farm. In such cases, list dwelling units that are actually located *within* the boundaries of the segment regardless of the location of the remainder of the farm. Do not list dwelling units located *outside* of the segment even though the rest of the farm on which they are located may lie within the segment boundaries.

RULE V. *Keep Strictly within the Boundaries of the Segment.*

Where the boundary line follows a street or road, it is intended to run down the middle of that street or road so that you list all dwelling units on the side which falls inside your segment but *not* those on the other side of the street or road segment. Where the segment boundary is a political boundary (county or city line) or the imaginary line, it is very important to list all units inside the boundary and no units outside it.

RULE VI. *Identify Each Unit Completely by Its Location.*

URBAN

a. *In urban areas list the address of each dwelling unit in full detail.*

Describe the unit in such a complete way that you or anyone else could find it in the future. Make sure that someone else could determine exactly which rooms are included in the unit and which are not included. If there is a house number, use it, plus any additional description which may be necessary. Where there are no numbers, describe the house as clearly as possible, e.g., "Spruce Street, large house, second from corner, red brick," "Spruce Street, near alley, small white house with uncovered porch," etc. If an unnumbered house is located next door to a numbered house, or between two numbered houses, identify the location of the unnumbered structure in relation to the nearest numbered structures–e.g., "Large red brick house between 1918 Spring Ave. and 1934 Spring Ave., immediately next door to 1918 Spring." While house colors should be used to improve your descriptions, do not depend on them alone to distinguish between houses–remember that houses can be and are repainted in other colors.

Complete street names should always be used. Do not leave out the words "street," "place," "boulevard," etc. When the same address applies on succeeding lines to several dwelling units, ditto marks should be used rather than writing the numbers or words over again. However, add to each unit sufficient identification to distinguish it clearly from all other units with the same or similar address. Do not leave the space blank in such case since this might be taken for an omission.

Always give apartment or room number in multi-dwelling structures. Where apartments are not numbered or lettered, write in floor and exact location of room or apartment (right, left, back, etc.). If more than one sleeping room on a floor is rented out to lodgers, describe each separate unit carefully, by giving its approximate location–e.g., "1519 Sherman Ave.–second floor, sleeping room on west side of building, directly north of stairway."

If a dwelling unit is in an alley with no name, describe the alley in some way that there can be no mistake in locating it in the future (e.g., "off 15th Street, between Buchanan and Johnson, running North-East.").

[For a sample of the proper way to list various types of *urban* dwelling units, see pages 160–165.]

RURAL

b. In rural areas, give address in full detail or a description of the house, its location, etc., so that anyone else can find it. Also show the location of the house on your map. In rural areas you may not find house numbers, and even road names may not be easily determined. The same general principle must be applied: Describe the location of the unit so that you or anyone else can find it. In many instances, *rural box numbers* appear on mail boxes, *as do actual names of box holders.* Both of these should be recorded wherever possible, but do not, in themselves, constitute an adequate address.

   It is also necessary that you *determine and record the rural route number* and indicate the Post Office address. Sometimes these route numbers may change, so it should not be assumed that all boxes on one stretch of road are on the same route. Those to a certain point may emanate from one city, and those beyond, from a different city or post office in the opposite direction. Also, in some areas, while all the mail boxes may be placed on one side of the road, some of them may describe dwelling units across the road and therefore possibly out of your segment.

   Dwelling units set back from the road must be checked by observation and inquiry, or else they may be overlooked. Record them following the most systematic manner possible. If such units have consecutively numbered mail boxes along the road, they are to be listed in that numbered sequence.

   Best description of a farm dwelling unit may be the *name of its owner or tenant*—"the John Miller farm." Since names change, however, it is best not to use them *alone* in the description. *Landmarks* should be employed where they will serve to describe adequately. A combination may be necessary—"Tenant's house with tin roof on Milller place."

   If a farm hand occupies a room in his employer's home, it is probable that he lives and eats with the farm owner, and thus does not occupy a separate dwelling unit of his own.

   [For a sample of the proper way to list various types of *rural* dwelling units see pages 166–169.]

DWELLING UNIT LISTING SHEET (DULS)

Segment No. 456-789

Page 1 **of** 4 **Pages**

County HONOLULU, HAWAII **Locality** Hauula

XXXXXX **ED** 10-11 **Name of PSU** HONOLULU, HAWAII

DIAGRAM OF SEGMENT

(Enter an "X" at your starting point and an arrow to show the direction of listing)

(Fill these boxes after completing the listing)
Lister's
Name John Doe
Date 12/64 Total DU's 40

List every DU located within this segment. Start at the most convenient point and proceed systematically aroun the segment. Go in and out of each street, alley, or road within the segment as you come to it. Continue until you have reached your starting point again. Use one line for each DU **BE SURE THAT EVERY PLACE OF RESIDENCE IS LISTED AND THAT ALL APARTMENTS, ETC ARE NUMBERED OR OTHERWISE IDENTIFIED S THAT SOMEONE ELSE COULD FIND THIS SPECIFIED DU.**

Line No.	**NAME AND ADDRESS OR DESCRIPTION OF DWELLING UNIT** Indicate street name and house number. Identify apartments by floor, number, and location. If this is not possible, describe DU and its location in detail. Also enter the name of the resident in this column for those DU's in which you can obtain names.	YEAR BUILT	RACE W or N	SURVEY NO.	S.
1	1 N. Main St. Bsmt. Apt.				
2	" " " " 3rd floor				
3	" " " " 2nd floor				
4	" " " " 3rd floor				
5	3 N. Main St. (1)				
6	7 N. Main St. Bsmt - front				
7	" " " " " rear				
.8	" " " " 3rd floor - right (2)				
9	" " " " " " left				
10	" " " " 2nd floor - right				
11	" " " " left				

(Use a continuation sheet, if necessary)

LISTING CONTINUATION SHEET

NORC
F 78
6/64

Segment No. 456-789

Page 2 of 4 Pages

County HONOLULU, HAWAII Locality Hauula

~~XXXXX~~ ED 10-11 Name of PSU HONOLULU, HAWAII

Line No.	NAME AND ADDRESS OR DESCRIPTION OF DWELLING UNIT Indicate street name and house number. Identify apartments by floor, number, letter, and location. If no street name or house number is available, describe DU and its location in detail.	YEAR BUILT	RACE W or N	SURVEY NO.	S.L.
12	9 N. Main St. Grd. floor Apt #1				
13	" " " " " " " #2				
14	" " " " 2nd floor " #101				
15	" " " " " " " #102				
16	11 N. Main St. Apt A				
17	" " " " " B				
18	1st DU just north of 11 N. Main (3) frame house with stone porch rail & chimney				
19	19 N. Main St. South side (or right side)				

20	" " " " Center entrance			
21	" " " " North side (or left side)			
22	21 N. Main St. Grd. floor - room - right front			
23	" " " " " " " " rear			
24	" " " " " " apt. left rear			
25	" " " " 2nd floor - room front right			
26	" " " " " " " " left			
27	" " " " " " " rear right			
28	" " " " " " " " left			
29	" " " " 3rd floor. apt			
30	23 N. Main St. Grd and 2nd floor (4)			
31	" " " " 2nd floor room entrance from outside rear stairs (4)			
32	21 N. Main St Grd. floor			
33	" " " " 2nd floor room front			

(Use another continuation sheet. if necessary)

NORC
F 78.
6/64

LISTING CONTINUATION SHEET

Segment No. 456-789

Page 3 of 4 Pages

County HONOLULU, HAWAII   Locality Hauula

XXXXX ED 10-11   Name of PSU HONOLULU, HAWAII

Line No.	NAME AND ADDRESS OR DESCRIPTION OF DWELLING UNIT Indicate street name and house number. Identify apartments by floor, number, letter, and location. If no street name or house number is available, describe DU and its location in detail.	YEAR BUILT	RACE W or N	SURVEY NO.	S.L.
34	27 N Main St. - 2nd fl. - room - 1st door from front DU				
35	" " " " - " " " 2nd door from front DU				
36	" " " " - " " " rear				
37	29 N Main St. left side of door left doorbell (5)				
38	" " " " left side of door right doorbell				
39	" " " " right side of door left doorbell				
40	" " " " right side of door right doorbell				

(1) Single family DU, requires no more information than this.

(2) Distinction here might be front and rear, rather than left and right.

(3) Where it is absolutely impossible to get a street number, get full description of DU and indicate specific location with reference to the nearest numbered DU.

(4) Example of listing of DU occupied by single family plus roomer (not related to family.) The room has a separate entrance and its occupant does not eat with the family occupying the rest of the house.

(5) Where obvious from number of door bells that there are several DU's in bldg. but you cannot ascertain their exact location, distinguish DU's by position of doorbells or by numbers assigned to door bells.

(Use another continuation sheet, if necessary)

NORC
F 78
6/64

DWELLING UNIT LISTING SHEET (DULS)

Segment No. 123-456

Page 1 of 2 Pages

County WASHOE, NEVADA — Locality Sparks

CT or ED 22-23 — Name of PSU RENO, NEVADA

DIAGRAM OF SEGMENT

(Enter an "X" at your starting point and an arrow to show the direction of listing)

(Fill these boxes after completing the listing)

Lister's Name John Doe

Date 11/64 Total DU's 10

List every DU located within this segment. Start at the most convenient point and proceed systematically around the segment. Go in and out of each street, alley, or road within the segment as you come to it. Continue until you have reached your starting point again. Use one line for each DU. **BE SURE THAT EVERY PLACE OF RESIDENCE IS LISTED AND THAT ALL APARTMENTS, ETC. ARE NUMBERED OR OTHERWISE IDENTIFIED SO THAT SOMEONE ELSE COULD FIND THIS SPECIFIED DU.**

Line No.	**NAME AND ADDRESS OR DESCRIPTION OF DWELLING UNIT** Indicate street name and house number. Identify apartments by floor, number, and location. If this is not possible, describe DU and its location in detail. Also enter the name of the resident in this column for those DU's in which you can obtain names.	YEAR BUILT	RACE W or N	SURVEY NO.	S. L.
1	1st DU on Rt. 73 north of intersection with Co. Hwy. 62 Large 2 story grey (1) frame house about 100 ft. north of Standard filling station (2) L.P. SMITH, Rt. 2, Sparks, Nev. (3)				
2	2nd DU on Rt. 73 north of Co. Hwy. 62 - 2 story white frame house with green trim with cupola on top. J.C. BROWN, Rt. 2, Sparks, Nev.				
3	Turning clockwise (to right) on Rt. 66, 1st DU east of I.C. RR - large 2 story white frame house with open porch across front - A.R. JONES, Rt. 2, Sparks, Nev.				
4	2nd DU east of RR on Rt. 66 - red house with white trim, weathered red barn and ruins of second barn. R.O. DOWNS, Rt. 3, Sparks, Nev.				
5	1st DU north of RR on N-S road running south from Rt. 66 white 2 story frame with green trim with barn on east side - P.L. BARNES, Rt. 3, Sparks, Nev.				
6	2nd DU north of RR on N-S road (south from Rt. 66) at dead end at top of hill - small grey frame house with windmill and barns to rear - H.B. CROWNE, Rt. 3, Sparks, Nev.				
7	Only DU on east side of N-S road (south from Rt. 66) south of RR tracks - faded yellow and brown R. McVEY, Rt. 3, Sparks, Nev.				

(Use continuation sheet, if necessary)

NORC
F 78
6/64

LISTING CONTINUATION SHEET

Segment No. 123-456

Page 2 of 2 Pages

County WASHOE, NEVADA Locality Sparks

XXXXXr ED 22-23 Name of PSU RENO, NEVADA

Line No.	NAME AND ADDRESS OR DESCRIPTION OF DWELLING UNIT Indicate street name and house number. Identify apartments by floor, number, letter, and location. If no street name or house number is available, describe DU and its location in detail.	YEAR BUILT	RACE W or N	SURVEY NO.	S
8	Back to Rt. 66 and turn to right on U.S. Hwy. 75 – 1st. DU on west side of Hwy. 75 about 200 ft. south of intersection of 66 and 75 – old stone house, chicken houses and barn to rear and right. H.S. WELLS Rt. 2, Sparks, Nev.				
9	Turn right off Hwy. 65 at 1st primitive road, go about ½ mile to 1st DU on right – red bungalow with green roof J. SMITH, Rt. 2, Sparks, Nev.				
10	2nd DU on right on primitive road (off Hwy 65) – large ranch-style house about 200 yds. off road – white frame and stone with TV aerial – T. Simmons, Rt. 2, Sparks, Nev.				
11					
12	(1) Mere description of color of DU is inadequate since DU may be repainted another color.				

13	(2) It is always helpful to indicate proximity to non-residential structure (such as filling station, post office, store, etc.)			
14	(3) Always include postal mailing address as well as rural route number.			
15				
16				
17				
18				
19				
20				
21				
22				

7.11 SUMMARY

The reader may be impressed or depressed by the difference in complexity and cost between the procedures in this chapter and those in Chapter 3. One solution is, of course, to use a multipurpose sample already in existence. The aim of multistage sampling is to select an efficient sample when lists are unavailable or only partly available. Given the great variance in the population sizes of individual clusters that lead to increased sampling error and inefficient workloads, the procedures described in this chapter are necessarily complicated.

The main focus of the chapter is on sampling pps (with probabilities proportionate to size). The use of sampling pps is demonstrated at several stages in the selection of an area sample as well as in other applications. It is shown that this procedure corrects for the variation in cluster sizes and yields an approximately optimum sample. Some variation remains because the cluster size measures used are imperfect, due to changes over time, or because, as with the use of a multipurpose sample, the size measure used is somewhat different than the actual one required. Even when cluster sizes vary considerably, sampling pps is not used if the entire cluster is selected.

Census sources of information for sampling pps are discussed briefly, primarily for samples of individuals or households. Procedures for combining census data and lists are given, along with the warning that the procedures should not overlap. Finally, instructions are given on how to obtain new lists for area sampling if this is necessary.

7.12 ADDITIONAL READING

Kish (40, ch. 6, 7, 9–11) is particularly strong on describing the detailed procedures for a multistage area sample. Some readers may be interested in noting the similarities between, and also the small differences in, the listing procedures used by the Survey Research Center, described by Kish and the National Opinion Research Center discussed here.

Hansen, Hurwitz, and Madow (35, Ch. 8, 9) give an excellent discussion of the effects of variation in cluster size and sampling pps. As in Chapter 1, however, the most useful sources for someone who is planning a multistage sample are descriptions of other multistage samples, such as *The Current Population Survey* (92), *The National Health Survey* conducted by the National Center for Health Statistics (94), as examples of large-scale government surveys; *Surveys of Consumers* (50) and other studies at the Survey Research Center, University of Michigan, such as *Justifying Violence* (7); *Volunteers for Learning* (39), *The Education of Catholic Americans* (34), and other NORC studies; and *The Gallup Poll* (28) and other state and regional surveys.

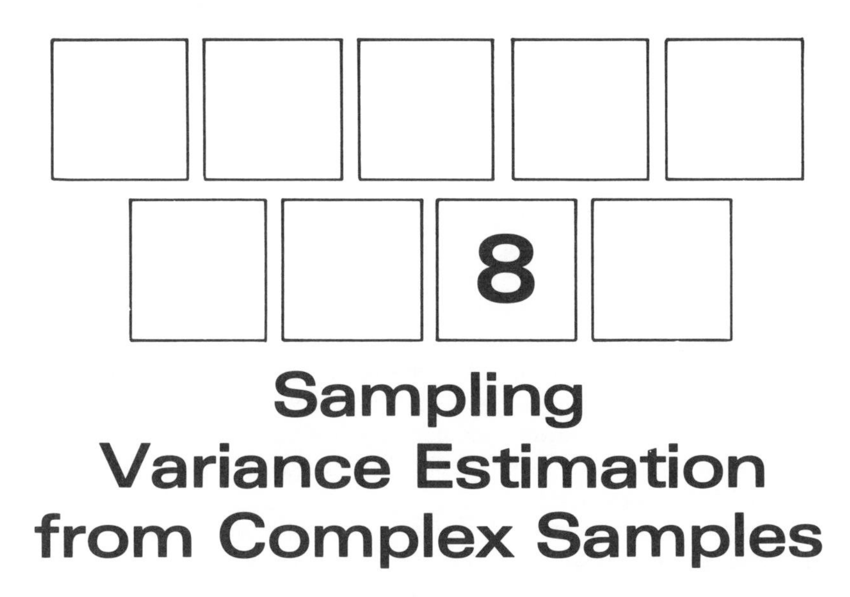

8 Sampling Variance Estimation from Complex Samples

An essential part of any careful sampling design is a procedure for estimating sampling variances of the results. Necessarily, as the sampling procedures become more complex, the procedures for estimating variance also become more complex. In this chapter are described the most efficient methods currently known for computing variances. The rationale for these methods is not difficult to understand. The actual computations are usually done by computer because there are so many repeated operations, although none of the formulas are very difficult. As with all parts of sample design, planning ahead pays off in efficient computation of variances. The researcher who waits until all the results have been tabulated and analyzed before thinking about variances usually will spend much more time and money on computations.

8.1 WHAT ARE SAMPLING VARIANCES?

Before turning to actual computational procedures, it is first necessary to define what is meant by a "sampling variance." It is unfortunate that the term "sampling error," or "standard error," has come to mean the same thing as the square root of sampling variance. Although, carefully used, error may suggest an inaccuracy where total accuracy is impossible, uninformed critics of a survey's findings sometimes seize on the term "sample error" to indicate carelessness or even an intent to deceive. Obviously, as used by statisticians, the word has no such meaning.

Perhaps even more dangerous is the use of the term "sampling error" by persons who have had only a smattering of statistics. Invariably, when they are asked what the term means, they reply that sampling error is the difference between the sample result and the "true" value in the population. Stated this strongly, sampling errors are defined to include all factors that might affect survey results–response effects due to questionnaire wording, interviewer effects, memory error, method of administration, and many other variables, as well as sample biases uncorrected by the Bayesian approach suggested in Chapter 6, errors in processing, and even errors in interpretation. If the issue were raised explicitly no one would claim that the measurement of sampling variances includes any of the factors just listed, but there is often the tendency to make this an implicit assumption.

The notion of a "true" value also is not very helpful. In the real world, the "true" value is never known: If it were, there would be no point to additional research. Even if a "true" value were known, there would be no guarantee that the results of any specific sample differed from that true value by exactly the sampling error.

The most meaningful and useful definition of a sampling variance involves the notion of repeated samples, all selected from the same population and using the same procedures. Then the sampling variance is defined as the variance between values of the statistic observed on these separate samples. The square root of this variance is the standard deviation (regretfully often referred to as the "standard error" or "sampling error").

The problem with this definition is that researchers seldom take repeated samples; usually, time and budget constraints limit the field work to a single sample. Thus, the task has been to devise procedures for computation that involve only a single sample. Before turning to such procedures, however, we present a famous example of the actual use of replicated samples for computing variances. If this procedure is possible, it has the advantage of being readily understood even by very unsophisticated users of survey data.

Example 8.1 Communism, Conformity, and Civil Liberties

In 1954, the Fund for the Republic sponsored a study, conducted by Samuel Stouffer (76) and his colleagues, to measure public attitudes toward civil liberties. (This study was conducted when the cold war and Senator Joseph McCarthy had made communism and civil liberties a major issue.) The study was very carefully designed and analyzed, and the report by Stouffer is a model of clarity. The book was intended to be used by decision makers and opinion leaders, as indicated by the final chapter, entitled "What the Answers to Some of these Questions Mean for People Who Can Do Something About Them."

A very large sample of almost 5000 interviews was completed by two highly qualified research organizations—the Gallup Poll and the National Opinion Research Center. Each organization drew its own sample and hired and trained its own interviewers, although the questionnaire was identical. Throughout the report, the results of the two organizations are compared as a measure of sampling variability. Where the results are very close, as in most cases, the reader is convinced that the data are reliable. Where there are large differences, the reader is alerted to the magnitude of sampling variance. Stouffer's (76) Table 1 is reproduced here as Table 8.1 because it exemplifies the effectiveness of this method of data presentation.

Table 8.1

Willingness to Give a Socialist the Right to Speak (Types of Community Leaders Compared with Cross-Sections)[a]

	Number of cases		Percentage willing to let socialist speak in their community		
	AIPO sample	NORC sample	AIPO sample	NORC sample	Combined sample
Public officials					
Mayors	56	56	79	80	80
Presidents, school board	53	58	90	85	88
Presidents, library board	56	53	93	91	92
Political party chairmen					
Republican county central committee	55	55	86	88	87
Democratic county central committee	53	50	96	87	91
Industrial leaders					
Presidents, chamber of commerce	57	56	82	88	85
Presidents, labor union	51	55	77	82	80
Heads of special patriotic groups					
Commanders, American Legion	55	59	81	70	76
Regents, D.A.R.	50	50	76	74	75
Others					
Chairmen, community chest	52	48	89	98	93
Presidents, bar association	53	46	94	96	95
Newspaper publishers	44	48	97	98	97
Presidents, women's club	54	55	62	75	68
Presidents, parent-teachers' association	53	59	83	75	79
Average of community leaders	742	758	84	84	84
Cross-section of same cities as leaders	409	488	60	61	61
National cross-section, rural and urban	2483	2450	57	60	58

[a]From Samuel Stouffer, *Communism, conformity, and civil liberties* (Garden City, New York: Doubleday, 1955), p. 31, Table 1. © 1955 by Doubleday & Company, Inc.

REPLICATION THROUGH SUBSAMPLING

Usually, the sample size is smaller than that of the example just given, and only a single survey organization is involved, but the same principle may be used. Whatever the sample size, the sample is split into subsamples for purposes of computing the variances, and these samples then are combined for purposes of estimating the parameters. Two questions arise: How is the sample split? and How many subsamples should be used?

PROCEDURES FOR SUBSAMPLING

If one has determined a total sample size n and a method of sampling, k subsamples are selected using the same procedure for each, and each having a sample size of n/k. If a multistage sample is selected with m primary sampling units, each of the subsamples will have m/k primary sampling units. Note that it is possible, using replication, that the same PSU will be selected in more than one of the subsamples.

The variance between these k subsamples is:

$$s^2 = \frac{\Sigma(x_i - \bar{x})^2}{k-1} \tag{8.1}$$

where x_i is the value of the statistic in the ith subsample.

What is wanted, however, is the variance between repeated samples of size n, not n/k. The variance for samples of size n is only $1/k$th as large as the variance in Formula (8.1), so the estimate for repeated samples of size n is

$$s^2 = \frac{\overset{k}{\Sigma}(x_i - \bar{x})^2}{k(k-1)}. \tag{8.2}$$

Example 8.2 Selection of Illinois Counties Using Replicated Sampling (Example 7.2 Continued)

Suppose one wishes to select a sample of 1000 households from Illinois, using the multistage procedures of the last chapter, and chooses to select 4 independent subsamples of 250 each, with Cook County and 5 other counties included in each subsample, so that the total sample, as before, is in Cook and 20 other counties. The sampling interval s for each of the subsamples is 1,736,103/5, or 347,220. Note that, unlike before, there are no other counties that fall into the subsamples with certainty except Cook County (see Table 7.2).

For each of the subsamples, a different random start is selected so that the selection numbers for the four subsamples are as shown in Table 8.3.

The selected counties are given in Table 8.2. Note that some counties are in two subsamples: both Rock Island and Kane Counties are in Subsamples 3 and

Table 8.2
Births, Deaths, and Marriages per 1000 Population for Counties in Four Selected Subsamples (1964)

Subsample	County	Births	Deaths	Marriages
1	Mason	18.6	14.7	7.9
	Livingston	18.0	10.3	6.9
	Macon	21.2	10.1	9.2
	Will	25.7	9.1	8.0
	DuPage	24.9	8.2	7.3
	Totals	21.7	10.5	7.9
2	Bond	18.2	12.8	6.1
	St. Clair	23.0	10.2	13.5
	Madison	22.5	9.2	10.0
	Winnebago	23.8	8.6	11.4
	Lake	24.3	7.8	14.3
	Totals	22.4	9.7	11.1
3	Franklin	13.5	16.1	10.0
	Randolph	17.2	12.3	6.7
	Stephenson	21.2	11.3	9.7
	Rock Island	22.3	10.2	12.8
	Kane	24.0	9.7	9.6
	Totals	19.6	11.9	9.8
4	Alexander	19.9	16.9	10.8
	Mercer	17.3	11.3	0
	LaSalle	19.6	11.0	7.8
	Rock Island	22.3	10.2	12.8
	Kane	24.0	9.7	9.6
	Totals	20.6	11.8	8.2

Table 8.3
Random Selection Numbers for Four Subsamples

	Subsample			
	1	2	3	4
r	295,801	193,180	25,231	6,001
$r+s$	643,021	540,400	372,451	353,221
$r+2s$	990,241	887,620	719,671	700,441
$r+3s$	1,337,461	1,234,840	1,066,891	1,047,661
$r+4s$	1,684,681	1,582,060	1,414,111	1,394,881

4. Within each county, a self-weighted sample is selected, as discussed in Chapter 7.

To illustrate the computation of sampling variances, an estimate of the number of births, deaths, and marriages per 1000 persons is also given for each county, based on sample results within the county. We assume that we need to estimate these parameters for the entire state, excluding Cook County, based only on the sample information.

The computations of the mean and sampling variance are as follows.

Births

$$\bar{x} = \frac{21.7 + 22.4 + 19.6 + 20.6}{4} = 21.1$$

$$s^2 = \frac{.6^2 + 1.3^2 + 1.5^2 + 1.5^2}{4(3)} = .38$$

Deaths

$$\bar{x} = \frac{10.5 + 9.7 + 11.9 + 11.8}{4} = 11.0$$

$$s^2 = \frac{.5^2 + 1.3^2 + .9^2 + .8^2}{4(3)} = .28$$

Marriages,

$$\bar{x} = \frac{7.9 + 11.1 + 9.8 + 8.2}{4} = 9.2$$

$$s^2 = \frac{1.3^2 + 1.9^2 + 0.6^2 + 1.0^2}{4(3)} = .56$$

HOW MANY SUBSAMPLES SHOULD BE USED

The decision of how many subsamples to use is difficult because there is no optimum number for all criteria. As more subsamples are used, the estimate of the sampling variance has greater precision. (Beginners often forget that every estimate based on a sample is subject to sampling variance and that the estimate of the variance is itself subject to variance.)

On the other hand, the number of subsamples is limited by the sample size, number of PSUs, and the desire to stratify the sample. The greater the number of subsamples, the less explicit or implicit stratification is possible. There is also the problem that, as the number of subsamples increases, the bias in the estimate of the mean also increases. Although this bias usually is very small, it may become important for estimates based on many replications.

Obviously, the minimum number of subsamples that can be used is two. Mahalanobis (48, 49) and Lahiri (43), who have used replicated procedures widely in India, generally use 4 replications. Deming (18, Ch. 11), using Tukey's method, suggests 10 replications. The use of 10 replications makes the computation of the sampling standard deviation particularly easy using the range, the difference between the largest and smallest values of the parameter.

For, between 3 and 13 replications, an approximate estimate of s is obtained from the relation:

$$s = \frac{\text{range}}{k} \tag{8.3}$$

For 10 replications, one merely observes the range and shifts the decimal point one place to the left to estimate s (see 40, p. 620). The same relation may be used for $k = 4$ or other values but requires some calculation.

Example 8.2 (Continued) Illinois Counties: Computing s Based on the Range

From Table 8.2, it is seen that the ranges and s are:

	Range	s_{Range}	s_{Direct}
Births:	22.4 − 19.6 = 2.8	.70	.62
Deaths:	11.9 − 9.7 = 2.2	.55	.53
Marriages:	11.1 − 7.9 = 3.2	.80	.75

The estimates of s, found by dividing these ranges by 4, are good but not identical estimates of the values of s found directly. The range would usually be used for quick evaluation of a few standard deviations since it is not an efficient procedure for use on a computer.

Although Kish (40) favors a large number of replications, it is difficult to think of applications for which this would be possible in the social sciences. My experience would suggest that, if this method is used, the number of subsamples should be limited to four. This makes sample selection, computation, and the visual presentation easier. The major objection to so few replications is that the estimates of sampling variance themselves are unreliable. This objection becomes less important if one is looking at not one but many statistics and variables, as is usually the case in the social sciences. Then, estimates of variance are usually averaged over similar statistics and variables to obtain a measure of design effect that is more reliable than estimates of variances of single statistics.

One of the most valuable uses of a split into two subsamples is in the search for a model–a procedure labeled "fishing" by Selvin and Stuart (74) in their article on data-dredging procedures in survey analysis. Typically, in fishing, one wishes to use multivariate procedures (either parametric or nonparametric) to explain some dependent variable. From a large list of possible independent variables, either the analyst or a computer program using stepwise regression methods selects some variables to include and others to exclude. The use of fishing procedures is very common, but it is not at all the same thing as testing a single hypothesis in advance. As Selvin and Stuart put it:

> We do not suggest that fishing is a reprehensible procedure–indeed it is often the only way to produce the food needed for the survey analyst's thought. We

> are arguing only that the survey analyst should admit that he *has* been fishing, rather than pretend that the model fell upon him as manna from heaven, and take the consequences, which are that the apparent probability levels of tests may have little relation to their true probability levels, and that the properties of estimates may be radically different from what is supposed [74, p. 22].

Since simple estimates of sampling variances provided by computer programs are meaningless when one is fishing, what can be done? The easy answer is to hold out half the data from the fishing process; one half of the sample is then used to generate hypotheses while the other half is used for testing. It is interesting to observe that many strong relations observed during the fishing phase vanish entirely on the test sample. The only drawback to this procedure is that it must be thought about before, rather than after, fishing.

8.2 PSEUDOREPLICATION

Samplers who would like to have their cake and eat it too have recently developed methods for computing variances from complex samples even if no subsamples are chosen. The procedure involves subsampling the data *after* they are collected. The subsampling must, as noted before, reflect the sampling procedures used. Thus, in a multistage sample, the sample selection is at the PSU level. It is assumed that the sample has been stratified and that two PSU selections have been made from each stratum. Half-samples are formed by selecting at random one of the two PSUs in a pair.

In the simplest application, only one replication is used. But, since replications can be generated by a computer program, it is possible to obtain many replications to increase the precision of the variance estimates. Even greater precision is possible by a procedure called "balanced replication," recently developed and used by McCarthy (95, 97) and by Kish and Frankel (41), based on a paper by Plackett and Burman (64). These procedures have general applicability regardless of the complexity of the sample design or the parameters studied, but do not, of course, correct for the fishing procedures discussed in the previous section.

Some recent investigations by McCarthy (95) suggest that balanced replication procedures are almost as reliable as doing all possible replications for means, and thus are highly efficient for computer application. Table 8.4 gives the balanced designs for 8, 12, 16, 20, and 24 pairs, as derived by Plackett and Burman (64), who also give larger balanced designs for up to 100 pairs. If the number of pairs is not a multiple of 4, the next largest design is used and the final columns omitted. Thus, for 10 pairs, the balanced design for 12 would be used, omitting the last two columns.

From each pair, one of the two PSUs is chosen at random to be "plus" while the other member of the pair is "minus." Thus, looking at Table 8.4, if there were eight pairs, the first replicate would consist of the PSUs designated "plus" in Pairs 1, 4, and 6, and "minus" in the other pairs. The last replicate would consist of all the "minus" PSUs.

A general program for computing variances, using balanced replication, is available from the Survey Research Center, University of Michigan—if it is not already in the library of programs for the computer the reader is using. It is efficient to identify PSU pairs in advance, but it is even more essential that each respondent be identified by PSU so that the data can be sorted easily. To reiterate the injunction at the beginning of this chapter—planning ahead pays off!

Table 8.4
Balanced Replication Designs for Computing Sampling Errors

N=8

Replication	Pair 1	2	3	4	5	6	7	8
1	+	-	-	+	-	+	+	-
2	+	+	-	-	+	-	+	-
3	+	+	+	-	-	+	-	-
4	-	+	+	+	-	-	+	-
5	+	-	+	+	+	-	-	-
6	-	+	-	+	+	+	-	-
7	-	-	+	-	+	+	+	-
8	-	-	-	-	-	-	-	-

N=12

Replication	1	2	3	4	5	6	7	8	9	10	11	12
1	+	-	+	-	-	-	+	+	+	-	+	-
2	+	+	-	+	-	-	-	+	+	+	-	-
3	-	+	+	-	+	-	-	-	+	+	+	-
4	+	-	+	+	-	+	-	-	-	+	+	-
5	+	+	-	+	+	-	+	-	-	-	+	-
6	+	+	+	-	+	+	-	+	-	-	-	-
7	-	+	+	+	-	+	+	-	+	-	-	-
8	-	-	+	+	+	-	+	+	-	+	-	-
9	-	-	-	+	+	+	-	+	+	-	+	-
10	+	-	-	-	+	+	+	-	+	+	-	-
11	-	+	-	-	-	+	+	+	-	+	+	-
12	-	-	-	-	-	-	-	-	-	-	-	-

(continued on next page)

Table 8.4 (continued)

N=16

Replication	Pair 1	2	3	4	5	6	7	8	9	10	11	12	13	14	15	16
1	+	-	-	-	+	-	-	+	+	-	+	-	+	+	+	-
2	+	+	-	-	-	+	-	-	+	+	-	+	-	+	+	-
3	+	+	+	-	-	-	+	-	-	+	+	-	+	-	+	-
4	+	+	+	+	-	-	-	+	-	-	+	+	-	+	-	-
5	-	+	+	+	+	-	-	-	+	-	-	+	+	-	+	-
6	+	-	+	+	+	+	-	-	-	+	-	-	+	+	-	-
7	-	+	-	+	+	+	+	-	-	-	+	-	-	+	+	-
8	+	-	+	-	+	+	+	+	-	-	-	+	-	-	+	-
9	+	+	-	+	-	+	+	+	+	-	-	-	+	-	-	-
10	-	+	+	-	+	-	+	+	+	+	-	-	-	+	-	-
11	-	-	+	+	-	+	-	+	+	+	+	-	-	-	+	-
12	+	-	-	+	+	-	+	-	+	+	+	+	-	-	-	-
13	-	+	-	-	+	+	-	+	-	+	+	+	+	-	-	-
14	-	-	+	-	-	+	+	-	+	-	+	+	+	+	-	-
15	-	-	-	+	-	-	+	+	-	+	-	+	+	+	+	-
16	-	-	-	-	-	-	-	-	-	-	-	-	-	-	-	-

N=20

Replication	Pair 1	2	3	4	5	6	7	8	9	10	11	12	13	14	15	16	17	18	19	20
1	+	-	+	+	-	-	-	-	+	-	+	-	+	+	+	+	-	-	+	-
2	+	+	-	+	+	-	-	-	-	+	-	+	-	+	+	+	+	-	-	-
3	-	+	+	-	+	+	-	-	-	-	+	-	+	-	+	+	+	+	-	-
4	-	-	+	+	-	+	+	-	-	-	-	+	-	+	-	+	+	+	+	-
5	+	-	-	+	+	-	+	+	-	-	-	-	+	-	+	-	+	+	+	-
6	+	+	-	-	+	+	-	+	+	-	-	-	-	+	-	+	-	+	+	-
7	+	+	+	-	-	+	+	-	+	+	-	-	-	-	+	-	+	-	+	-
8	+	+	+	+	-	-	+	+	-	+	+	-	-	-	-	+	-	+	-	-
9	-	+	+	+	+	-	-	+	+	-	+	+	-	-	-	-	+	-	+	-
10	+	-	+	+	+	+	-	-	+	+	-	+	+	-	-	-	-	+	-	-
11	-	+	-	+	+	+	+	-	-	+	+	-	+	+	-	-	-	-	+	-
12	+	-	+	-	+	+	+	+	-	-	+	+	-	+	+	-	-	-	-	-
13	-	+	-	+	-	+	+	+	+	-	-	+	+	-	+	+	-	-	-	-
14	-	-	+	-	+	-	+	+	+	+	-	-	+	+	-	+	+	-	-	-
15	-	-	-	+	-	+	-	+	+	+	+	-	-	+	+	-	+	+	-	-
16	-	-	-	-	+	-	+	-	+	+	+	+	-	-	+	+	-	+	+	-
17	+	-	-	-	-	+	-	+	-	+	+	+	+	-	-	+	+	-	+	-
18	+	+	-	-	-	-	+	-	+	-	+	+	+	+	-	-	+	+	-	-
19	-	+	+	-	-	-	-	+	-	+	-	+	+	+	+	-	-	+	+	-
20	-	-	-	-	-	-	-	-	-	-	-	-	-	-	-	-	-	-	-	-

(continued on next page)

Table 8.4 (continued)

N=24	Pair																						
Replication	1	2	3	4	5	6	7	8	9	10	11	12	13	14	15	16	17	18	19	20	21	22	23	24
1	+	-	-	-	-	+	-	+	-	-	+	+	-	-	+	+	-	+	-	+	+	+	+	-
2	+	+	-	-	-	-	+	-	+	-	-	+	+	-	-	+	+	-	+	-	+	+	+	-
3	+	+	+	-	-	-	-	+	-	+	-	-	+	+	-	-	+	+	-	+	-	+	+	-
4	+	+	+	+	-	-	-	-	+	-	+	-	-	+	+	-	-	+	+	-	+	-	+	-
5	+	+	+	+	+	-	-	-	-	+	-	+	-	-	+	+	-	-	+	+	-	+	-	-
6	-	+	+	+	+	+	-	-	-	-	+	-	+	-	-	+	+	-	-	+	+	-	+	-
7	+	-	+	+	+	+	+	-	-	-	-	+	-	+	-	-	+	+	-	-	+	+	-	-
8	-	+	-	+	+	+	+	+	-	-	-	-	+	-	+	-	-	+	+	-	-	+	+	-
9	+	-	+	-	+	+	+	+	+	-	-	-	-	+	-	+	-	-	+	+	-	-	+	-
10	+	+	-	+	-	+	+	+	+	+	-	-	-	-	+	-	+	-	-	+	+	-	-	-
11	-	+	+	-	+	-	+	+	+	+	+	-	-	-	-	+	-	+	-	-	+	+	-	-
12	-	-	+	+	-	+	-	+	+	+	+	+	-	-	-	-	+	-	+	-	-	+	+	-
13	+	-	-	+	+	-	+	-	+	+	+	+	+	-	-	-	-	+	-	+	-	-	+	-
14	+	+	-	-	+	+	-	+	-	+	+	+	+	+	-	-	-	-	+	-	+	-	-	-
15	-	+	+	-	-	+	+	-	+	-	+	+	+	+	+	-	-	-	-	+	-	+	-	-
16	-	-	+	+	-	-	+	+	-	+	-	+	+	+	+	+	-	-	-	-	+	-	+	-
17	+	-	-	+	+	-	-	+	+	-	+	-	+	+	+	+	+	-	-	-	-	+	-	-
18	-	+	-	-	+	+	-	-	+	+	-	+	-	+	+	+	+	+	-	-	-	-	+	-
19	+	-	+	-	-	+	+	-	-	+	+	-	+	-	+	+	+	+	+	-	-	-	-	-
20	-	+	-	+	-	-	+	+	-	-	+	+	-	+	-	+	+	+	+	+	-	-	-	-
21	-	-	+	-	+	-	-	+	+	-	-	+	+	-	+	-	+	+	+	+	+	-	-	-
22	-	-	-	+	-	+	-	-	+	+	-	-	+	+	-	+	-	+	+	+	+	+	-	-
23	-	-	-	-	+	-	+	-	-	+	+	-	-	+	+	-	+	-	+	+	+	+	+	-
24	-	-	-	-	-	-	-	-	-	-	-	-	-	-	-	-	-	-	-	-	-	-	-	-

Following the notation of Kish and Frankel (41):

B is the population value being estimated, neglecting differences from some true value B_{true}, due to measurement errors, nonresponse, etc.

$\hat{b}$ is the statistic used by the researcher for estimating B, whose variance var($\hat{b}$) needs to be estimated; we neglect the possible existence of some better estimator b^*.

$\bar{b}_j = (b_j + b_j')/2$ is the mean of a replication and its complement.

The complement of any replicate consists of all the remaining PSUs not included in the replicate.

$\text{var}(\bar{b}_j) = (b_j - b_j')^2/4$ denotes the estimate of the variance of $\bar{b}_j$.
$\overline{\text{var}}_k(\bar{b}_j)$ denotes estimates of the variance of $\bar{b}_j$, averaged over k values.

8.3 VARIANCE ESTIMATES FROM SYSTEMATIC SAMPLES WITH IMPLICIT STRATIFICATION

The balanced replication procedure just described assumed that the sample was stratified and that two PSUs had been sampled in each stratum. In many multistage samples, the PSUs are sampled systematically after being arranged in sequence by geography, size, or other stratifying variables. This yields an implicit stratification, but one in which there is only one selection per stratum. There is no *exact* way to compute variances in this case.

The standard procedure used, however, is to combine two adjacent selections into a stratum and proceed as in the previous section. Noncertainty PSUs are paired with adjacent PSUs, and certainty PSUs are split into two equal halves by location and block, these halves being considered as the two selections from the stratum.

This method yields an estimate of the variance that is known to be slightly biased upward. Many samplers, however, are willing to use this method, since it provides a possible increase in sampling efficiency, even at the cost of a slight overestimation of variance. A good discussion of this procedure is given by Kish (40, Sec. 8.6B).

Example 8.3 Computing Sampling Variances Using Balanced Replication (Example 7.2 Continued)

For this example, we use the initial 20 Illinois counties selected in Example 7.2 and designate successive groups of two as pairs. Using a table of random numbers, the first of each pair is designated as "+" if an odd number is observed in the table and is designated as "–" if an even number is observed. The other member of the pair is then given the complementary designation. Table 8.5 provides the basic information on births, deaths, and marriages. All the "+" PSUs are listed first, for ease in following the example.

The estimates of sampling variance are computed by using the balanced design for N=12, omitting the last two columns. The computations are given in Table 8.6. Note that there is substantial variability between the replications, but the final estimate of variance is highly reliable.

Some readers may wonder at the difference in the estimates of the variance given in Examples 8.2 and 8.3. All the estimates using balanced replication are

Table 8.5
Births, Deaths, and Marriages per 1000 Population
for Selected Counties (1964)

County	Pair		Births	Deaths	Marriages
Pike	1	+	16.6	12.9	8.8
Williamson		–	17.5	14.2	8.8
Clark	2	+	15.2	13.9	7.9
Fulton		–	17.8	13.5	8.3
Coles	3	+	17.8	10.4	8.7
Vermilion		–	19.4	10.4	10.1
St. Clair	4	+	23.0	10.2	13.5
LaSalle		–	19.6	11.0	7.8
Champaign	5	+	19.0	6.5	8.3
Madison		–	22.5	9.2	10.0
Sangamon	6	+	23.0	12.0	10.2
Kankakee		–	21.5	9.0	9.3
Peoria	7	+	20.8	10.0	9.0
Tazewell		–	23.0	8.0	7.4
Winnebago	8	+	23.8	8.6	11.4
Will		–	25.7	9.1	8.0
McHenry	9	+	24.5	9.4	9.2
Lake		–	24.3	7.8	14.3
$DuPage_1$	10	+	24.9	8.2	7.3
$DuPage_2$		–	24.9	8.2	7.3
Totals			21.2	10.1	9.3

Table 8.6
Sampling Variance Computations
Using Balanced Replications

	Births		Deaths		Marriages	
Replications	$\hat{b}$	$\text{var}(\bar{b}_j)$	$\hat{b}$	$\text{var}(\bar{b}_j)$	$\hat{b}$	$\text{var}(\bar{b}_j)$
1	21.0	.04	10.2	.01	9.0	.09
2	21.4	.04	10.0	.01	9.5	.04
3	20.9	.09	10.0	.01	8.3	1.00
4	21.9	.49	10.1	0	9.6	.09
5	21.0	.04	9.8	.09	9.6	.09
6	20.7	.25	9.9	.04	9.2	.01
7	21.5	.09	10.7	.36	9.3	0
8	21.0	.04	9.8	.09	9.9	.36
9	21.6	.16	10.1	0	9.4	.01
10	21.1	.01	10.3	.04	8.7	.36
11	21.1	.01	10.5	.16	9.7	.16
12	21.6	.16	10.0	.01	9.1	.04
Average $\text{var}_k(\bar{b}_j)$		.12		.07		.19

smaller than the estimates using actual replicated samples, and, for births and deaths, the differences are substantial. This is because the pseudoreplication procedures allow for greater use of stratification and thus for a more efficient sample. Looking at the county data, it is evident that both births and deaths are functions of the age distribution of the population. The implicit stratification by median income evidently also controls for age. To generalize—whenever one expects stratification to improve the efficiency of the sample, it is better to get the maximum effects of stratification by using a single sample and balanced replication rather than by using actual replicated samples.

8.4 MEASURING DESIGN EFFECTS

As discussed in Chapter 4, it often is useful to have measures of the design effects for a study either to estimate sampling variances for parameters for which the estimates were not computed directly, or for use in planning future studies. In measuring design effects, one computes the ratio of the sample variance of a complex sample, as found by the methods just discussed, to the variance of a simple random sample of the same size. Although this estimate will vary for different variables, due to sampling variability as well as differences in intracluster correlations and stratification effects, it is generally possible to average estimates of design effects for similar variables.

For a simple random sample of size n when selected units are not replaced, the variance of a statistic $\hat{b}$ is computed by:

$$s_{\hat{b}}^2 = \left(\frac{N-n}{N-1}\right)\frac{s^2}{n}. \tag{8.4}$$

THE FINITE CORRECTION FACTOR

The term $(N-n/N-1)$ is called the "finite correction factor." It is evident that, if all the elements in the population are sampled, there is no sampling error; this is reflected in the finite correction factor, which becomes zero if $n = N$. In almost all cases, N will be large enough so that the difference between N and $N-1$ is trivial. Then the finite correction factor becomes $N-n/N$, or $(1-f)$, where f is the sampling fraction n/N.

For simplicity, f itself usually is ignored if it is less than .01 or .02. In most general population samples, f is far less than .01 and may be safely ignored; for special groups and non-human populations, f is frequently large and usually should be included in the variance computations. There is an exception. A small population may be considered as part of some superpopulation through time, about which one is really concerned. Thus, in a study of graduate education, one has only available students at a given point in time, but may really be interested in predicting the behavior of students in the next decade or century. Similarly, in studies of organizational behavior, firms are observed during given time periods when one really wishes to make statements about the superpopulation of firms in the future. In these cases, the finite correction factor becomes zero, since N becomes infinite. (Studies at one point in time also introduce unknown sample biases that are beyond the scope of this chapter to discuss.)

Example 8.4 Measuring Design Effects

The design effects reported by Kish and Frankel (41) are summarized in Table 8.7. Based on the Survey Research Center sample of 12 certainty PSUs and 62 other PSUs, 47 strata were developed for computing. (The 62 noncertainty PSUs were split into 31 strata, and the certainty PSUs were split into 16 strata, combining by census tracts and blocks.) The parameters are primarily regression coefficients, and the results suggest much smaller design effects for regression coefficients than for means and proportions.

Kish and Frankel also report on a national health examination survey conducted by the National Center for Health Statistics. In this study, the design effects are larger because of heavier clustering (there were only 42 PSUs) and because examinations in each PSU were conducted by the same medical team. Again, however, the regression coefficients are subject to smaller design effects than are means.

Table 8.7
Design Effects on Various Surveys

Study	Sample size	Parameters	Average design effect
Private pensions and savings	1853	Regression coefficients	1.12[a]
Voting	1111	Regression coefficients	1.03[b]
		Simple correlation	1.20
		Partial correlation	1.08
		Mean	1.22
Social class	3990	Dummy variable	1.21[c]
		Regression coefficients	
Health examination	3091	Ratio means	3.08[d]
		Regression coefficients	1.60
		Simple correlation	1.64
		Partial correlation	1.82
		Multiple R	2.02

[a]From Kish and Frankel (41, Table 1).
[b]From Kish and Frankel (41, Table 2).
[c]From Kish and Frankel (41, Table 4).
[d]From Kish and Frankel (41, Table 3).

8.5 VARIANCE COMPUTATION IN SPECIAL CASES

The procedure just described is applicable to all kinds of sample designs and variables, but is particularly appropriate when both clustering and stratification are used and when variables other than simple means and proportions are estimated, so that simple error formulas do not exist or would be very costly to compute. If a simple random or systematic sample is used, one may use the simple random sample variance formula

$$s_{\bar{x}}^2 = \frac{(1-f)s^2}{n} = \frac{(1-f)[\sum_{i=1}^{n} x_i^2 - n\bar{x}^2]}{n(n-1)} \tag{8.5}$$

with the finite correction if needed. This formula is *not* appropriate for the case in which a simple random sample is selected from a cluster that is chosen judgementally or for convenience. If a city like Cleveland is selected to represent the entire United States, no estimates of variance are meaningful, regardless of how the sample is selected within that city. Similarly, the use of variance estimates is meaningless if students are selected at random from convenient school classrooms.

ULTIMATE CLUSTER VARIANCE ESTIMATES

If a multistage self-weighting sample is used but there is no stratification, the PSU totals may be considered as elements of a simple random sample and $s_{\bar{x}}^2$ may be estimated by

$$s_{\bar{x}}^2 = (1-f)\frac{\sum^{m} (x_i - \bar{x})^2}{m(m-1)} \tag{8.6}$$

where x_i is the total for the statistic in the ith PSU summed over all the sample elements in that cluster, and $\bar{x}$ is the average taken over all clusters.

$$\bar{x} = \frac{\sum^{m} x_i}{m}$$

This formula is appropriate only for means or proportions and *when all the sample PSUs are about the same size.* Thus, this method is not directly usable if some PSUs are selected with certainty. If there is any explicit or implicit stratification, Formula (8.6) ignores it and, thus, will overestimate the variance. As may be seen, this formula uses the same input, PSU totals, as does balanced replication, but the PSUs are not paired. Formula (8.6) is simpler and less costly than balanced replication, but it has much less general application.

In a slightly more complex and general situation, one is not interested in the mean but in some estimate of a ratio $r = y/x$, where it is possible that x could be the sample size if unequal sample sizes are selected within cluster. Then an approximate formula for the variance of r in the ultimate clusters is:

$$s^2(r) = \frac{1-f}{x^2}\,\frac{m}{m-1}\left[\left(\sum^m y_i^2 - \frac{y^2}{m}\right) + r^2\left(\sum^m x_i^2 - \frac{x^2}{m}\right) - 2r\left(\sum^m y_i x_i - \frac{yx}{m}\right)\right] \quad (8.7)$$

where

$$r = \frac{y}{x} = \frac{\sum^m y_i}{\sum^m x_i}$$

It is, of course, possible to use the balanced replication procedures discussed earlier to compute the variance in this case; as the formulas become more complex, balanced replication becomes essential.

Example 8.5 Ultimate Cluster Variance Estimates (Example 7.2 Continued)

The data from Table 8.5 may be used to illustrate that Formula (8.6) overstates the sampling variance when stratification has been used. Using Formula (8.6) but ignoring the finite correction factor, since f is near zero, and assuming that cluster sample sizes are equal, the estimated values of $s_{\bar{x}}^2$ are:

Births	.50
Deaths	.24
Marriages	.18

These values are considerably higher than those in Table 8.6, but are approximately the same as those given in Example 8.2, in which four subsamples were actually selected. This is again a result of the reduction in variance because of the implicit stratification of counties. Only the balanced replication method allows for, and measures, this stratification.

Using the same data, suppose we are interested in $r = y/x$, where y represents births and x represents deaths. Then

$$\sum^m y_i = 424.8 \qquad \sum^m x_i = 202.5$$

$$\sum^m y_i^2 = 9213.88 \qquad \sum^m x_i^2 = 2141.21$$

$$\sum^m y_i x_i = 4210.05 \qquad m = 20$$

$$r = \frac{y}{x} = \frac{424.8}{202.5} = 2.10$$

$$s^2(r) = \frac{1}{(202.5)^2} \cdot \frac{20}{19} [191.13 + 4.41(90.90) - 4.2(-91.05)]$$

$$= .025 \text{ (ignoring the finite correction term)}$$

VARIANCE ESTIMATES FOR STRATA OR SUBSAMPLES

The use of replication for variance estimates is, of course, not limited to the total sample. The same procedures may also be used for computing the variance of any subsample or stratum from the initial sample. For example, if separate estimates of variances by race are wanted, the sample is first split by race; then, for blacks and whites separately, the data are sorted into pairs of PSUs, and balanced replication is used. From these estimates of strata variances, it is also possible to compute total sample variances using Formula (6.10), but, unless there are only a few strata, it is usually easier to recombine the strata before computing total sample variances.

Since the estimate of the sample variance itself is subject to variance, which is a function of sample size, the estimates of sample variances for strata will be less reliable than the estimate of sampling variance for the total sample.

8.6 COMBINING VARIANCE ESTIMATES WITH PRIOR INFORMATION

The Bayesian procedures discussed in Chapters 5 and 6 assume simple random sampling or at least simple random sampling within stratum. The generalization to complex samples is direct.

a. Prior Information about Total Population. If one has information about the total population which can be expressed as a normal prior distribution with mean μ_{prior} and variance V^2, and if, after sampling, one estimates the sampling variance as $s_{\hat{b}}^2$, then the posterior variance is estimated from:

$$\frac{1}{V_{\text{post}}^2} = \frac{1}{V_{\text{prior}}^2} + \frac{1}{s_{\hat{b}}^2} \qquad (8.8)$$

That is, the reciprocal of the posterior variance is the sum of the reciprocals of the variances of the prior and sampling distributions. This is the same as computing the variance when two independent samples are combined.

b. Differential Information by Stratum. If, as in Chapter 6, there is differential information by stratum, it is necessary to compute variance estimates for each stratum separately from the sample data. These sample estimates are then combined for each stratum with prior information, using Formula (8.8). Finally, the strata are combined by using Formula (6.10) to give an estimate for the total population.

In many cases, unless the data user has very strong prior beliefs, the sample results will dominate, especially if a large sample has been selected (say, several hundred or more). In this situation, the Bayesian variance estimates will be only slightly smaller than the sampling variance estimates.

Many social scientists or granting agencies may use explicit or implicit Bayesian analysis in deciding what studies to conduct or to fund and how large the sample should be. If the results are intended for publication in a professional journal having a wide audience, the sampling variances should be presented so that individual readers with varying priors may individually compute their posterior variances if they wish to do so. If the author has strong prior beliefs, he may also wish to present his posterior distribution to the reader.

8.7 SUMMARY

Sampling variances are best defined as the differences in a sample statistic that are observed when repeated samples are selected from the same population, using the same procedures. Since actual replication is uncommon, due to time and budget constraints, replication is achieved through subsampling or by the use of pseudoreplication procedures after the data are collected. If multistage sampling is used, the sample is split by PSUs.

Actual subsampling reduces the degree of stratification possible and, thus, may reduce sample efficiency. For this reason, pseudoreplication procedures involving balanced replication appear to be most useful. These procedures group sample PSUs into pairs and select half-samples by choosing one PSU from each pair. Since individual replications are subject to high variability in their estimates of variance, repeated replications by computer are required for reliability. Balanced repetitions, using orthogonal designs, are found to be highly efficient. (Balanced designs are given in the chapter for 8, 12, 16, 20, and 24 pairs of PSUs.)

Formulas for computing simple random sampling variances have little general applicability but are often used to compute the design effect, the ratio of the actual sampling variance to the variance of a simple random sample of the same size. Finally, there is a brief discussion of procedures for combining variance estimates from complex samples with prior information, using Bayesian analysis.

8.8 ADDITIONAL READING

The use of replicated samples was first suggested about three decades ago by Mahalanobis (48, 49), whose papers are straightforward and easy to follow. In the United States, Deming has written extensively on replicated sampling, especially in *Sample Design in Business Research* (18).

The ideas about pseudoreplication are much more recent, and work in this area is still continuing. The papers, cited in this chapter, by McCarthy (95, 97) and Kish and Frankel (41) are not only comprehensive but also comprehensible to most readers of this book. The standard texts are not very useful for this topic, excepting Kish (40, Ch. 6), who describes his earlier work on estimation of variances for complex samples. An example of the use of pseudoreplication in program evaluation research is given by Finifter in *Sociological Methodology:1972* (25).

Another form of pseudoreplication is jack-knife sampling, first discussed by John Tukey. A relatively easy discussion of jack-knife sampling is given by Brillinger (9) in *Commentary,* the Journal of the British Market Research Society.

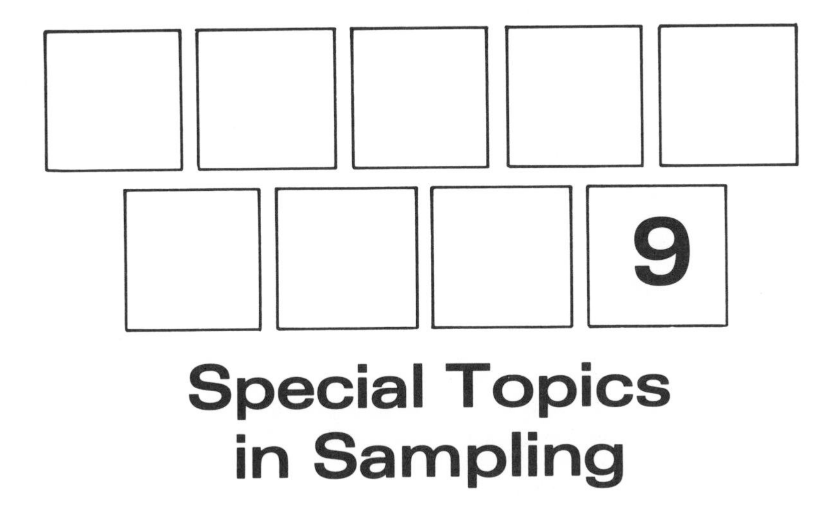

9 Special Topics in Sampling

There are important topics in sampling not covered in the earlier chapters. Although it is impossible to discuss all of them here, four have been selected because they have been of particular interest to me and, I think, will be of interest to many social scientists.

The first topic deals with methods for reducing the cost of data collection by substituting quota procedures for call-backs. The second deals with methods for locating rare special populations, the third with snowball sampling, and the last with the special procedures for the continuous sampling of panels.

9.1 PROBABILITY SAMPLING WITH QUOTAS

Three decades ago, when the advocates of probability sampling met and defeated the defenders of quota sampling, the doctrine became established that there is an unbridgeable gulf between the two methods. It was conceded that quota samples were cheaper, but most sampling statisticians had no doubts that quota samples were far less accurate than probability samples and that, even worse, there was no way to measure the accuracy of a quota sample.[1]

[1] Illustrations of the typical views toward quota sampling in earlier texts are found in Hansen, Hurwitz, and Madow (35, Vol. I, p. 71):

> The so-called "quota controlled" sampling method, which has been widely used, is essentially a sample of convenience but with certain controls imposed

Since then, there has been a major change in quota sampling methods, particularly since the failure of the polls in 1948 (63). The major change has been the establishment of tight geographical controls that the interviewer must follow. That is, in her search to fill quotas, the interviewer follows a specified travel pattern, visiting predesignated households.

It is the intent of this section to attempt a rationalization of this procedure, which indicates that it is very close to traditional probability sampling. To

---

> that are intended to avoid some of the more serious biases involved in taking those most conveniently available.
>
> The restrictions imposed on the convenience of the interviewer by this method may possibly considerably reduce the biases. However, they may also be completely ineffective. What is worse, there is no way to determine the biases except by a sample properly drawn and executed.

In William Cochran (14, p. 105), a similar, but slightly more favorable view is taken of quota sampling:

> Another method that is used in this situation (stratified sampling where the strata cannot be identified in advance) is to decide in advance the n_h that are wanted from each stratum and to instruct the enumerator to continue sampling until the necessary "quota" has been obtained in each stratum. If the enumerator initially chooses units at random, rejecting those that are not needed, this method is equivalent to stratified random sampling. . . . As this method is used in practice by a number of agencies, the enumerator does not select units at random. Instead, he takes advantage of any information which enables the quota to be filled quickly (such as that rich people seldom live in slums). The object is to gain the benefits of stratification without the high field costs that might be incurred in an attempt to select units at random. Varying amounts of latitude are permitted to the enumerators. . . . Sampling theory cannot be applied to quota methods which contain no element of probability sampling. Information about the precision of such methods is obtained only when a comparison is possible with a census or with another sample for which confidence limits can be computed.

According to W. Edwards Deming (18, p. 31):

> There is another kind of judgment sample called a quota sample. The instructions in a quota sample ask the interviewers to talk to a specified number of people of each sex and age, perhaps by section of the city, perhaps by economic level. The report of the results usually boasts of good agreement between the sample and the census in respect to the classes specified, but what does this mean? It means that the interviewers reported what they were supposed to report concerning these classes; it proves little or nothing with respect to the accuracy of the data that constitute the purpose of the study. . . . There is no way to compare the cost of a probability sample with the cost of a judgment sample, because the two types of sample are used for different purposes. Cost has no meaning without a measure of quality, and there is no way to appraise objectively the quality of a judgment sample as there is with a probability sample.

differentiate it from older quota sampling methods that do not specify a travel pattern, the procedure will be referred to as "probability sampling with quotas." This procedure is not unbiased, but typically the bias is small. On the other hand, a careful cost analysis indicates that differences in direct interviewer costs between probability sampling with call-backs and probability sampling with quotas is also small. The major advantage of this new procedure may well be the speed with which interviewing can be completed. Thus, when speed is critical to obtain immediate public reaction to a crisis, such as the Kennedy assassination, probability sampling with quotas can be most useful. The National Opinion Research Center completed the field work on a national study of public reactions to the president's assassination in about 10 days, using a probability sample with quotas. Quota studies with less urgency are finished in 2 or 3 weeks. On the other hand, regular probability samples usually take 6 weeks or longer.

9.2 ASSUMPTIONS UNDERLYING PROBABILITY SAMPLING WITH QUOTAS

In probability sampling with call-backs, the interviewer is given a specific household or individual to interview. If the individual is not available on the first call, repeat call-backs are made until the interview is obtained or the respondent refuses to grant one.

In probability sampling with quotas, the basic assumption is that it is possible to divide the respondents into strata in which the probability of being available for interviewing is known and is the same for all individuals within the stratum, although varying between strata. Any respondent's probability of being interviewed is the product of his initial selection probability times his probability of being available for interviewing. Although these probabilities will not be identical for all respondents, they can be determined, and the sample is therefore a probability sample. There is an implicit assumption that an interviewer in a sample segment follows the same time pattern over repeated surveys and that the respondent has a pattern of availability depending on certain characteristics.

The quotas then used must clearly be associated with the probability of being available for interviewing. Essentially, the quotas should be based on the reciprocals of the probabilities of availability. If the probability of the individuals in Stratum A being available is twice as large as the probability of individuals in Stratum B, the sampling rate for Stratum A should be one-half that for Stratum B.

In the usual situation, quotas are set for a given stratum on the basis of the sampling rate and universe estimates of the stratum size. These quotas are normally determined for the smallest geographic area for which information is available. Thus, in metropolitan areas, census tract information is used, while in nontracted areas, the quotas are based on the characteristics of the locality or of the rural portion of the county. This method introduces the possibility of error because of inadequate universe estimates, but generally it is almost like the method that uses sampling rates directly.

This procedure is wasteful from a sampling viewpoint because households with no one at home are skipped, along with households in which the respondent is not available for interviewing at the time the interviewer calls, or households with respondents who do not fit the quota. The field-cost savings, however, considerably exceed the increase in internal sampling costs.

Probability sampling with quotas has been used primarily for sampling of *individual respondents.* Where household behavior or opinions are wanted, it would be possible to use the same procedure, but, since size of household is highly correlated with availability, it would be necessary to make it a major quota control. Because any knowledgeable adult is acceptable as the respondent in a household survey, probability sampling with call-backs of households is less costly than sampling of designated respondents in households. Generally, cost and time savings or probability samples with quotas of households will not be great enough to make this method very useful, considering the possible biases.

The rationalization of probability sampling with quotas depends on the major assumption that respondents may be stratified by their probabilities of being available for interviewing, while probability sampling with call-backs does not require this assumption. Fortunately, there is strong evidence that this assumption is *almost* true for the kinds of surveys generally conducted in the United States. To the extent that the assumption is not true, small biases are introduced, but the method still remains a probability sample.

Even in the usual probability sample with call-backs, biases exist because of noncooperators. These same biases exist in probability sampling with quotas. We have not observed any major difference in the overall cooperation rates achieved by interviewers on probability samples with call-backs as compared to probability samples with quotas. These cooperation rates depend on both respondents and interviewers. Since respondents cannot be aware of the type of sampling, any difference must be due to the fact that the interviewer did not try as hard as she might have to convert a refusal into a cooperator. When interviewers do both types of interviews, there is no evidence that this is occurring.

It may be useful for the reader to compare the rationale for the samples just described with that of the Politz–Simmons (65) weighting method sometimes used to adjust for not-at-home bias. In this procedure, no call-backs are made and no quotas are used. Typically, the respondent is asked whether or not he was home on the preceding five nights, and his answers to this question determine the weight he receives. Thus, a respondent who had been home all nights is assigned a weight of 1, and a respondent who had not been home on any of the preceding five nights gets a weight of 6, since only one-sixth of the respondents of this type would be found at home on a random night.

The Politz–Simmons weighting has three disadvantages. First, the weighting depends on the respondent's memory of how he spent the last five nights, and, in general, respondents will tend to overstate their availability. Second, the use of weights increases the sampling variability substantially. Finally, the weighting method requires careful controls when the data are processed, to ensure that the weighting is done properly. It is my impression that not very many surveys currently use the Politz–Simmons weighting procedure, because of these difficulties.

If one were willing to accept the answer to the at-home question as being reliable and did not worry about the cost of weighting, it would be possible to develop a combined sampling method that uses probability sampling with quotas to keep sampling variability low, and uses the answer to the at-home question to eliminate remaining sample biases.

9.3 RESPONDENT CHARACTERISTICS RELATED TO AVAILABILITY FOR INTERVIEWING

How does one go about establishing strata within which individuals have equal probability of being available for interviewing, and how are these strata tested for homogeneity? Since direct data are unavailable, one must use past experience with probability samples. Many earlier studies have shown that women generally are more readily available for interviewing than are men. Primarily, this is because more men than women are employed. When one imposes the additional control of employment status, one sees a substantial difference between employed and unemployed women, but the difference between men and women shrinks. In addition, age of men is of some importance.

Table 9.1 gives the average number of calls required to complete an interview—by age, sex, and employment status—for typical probability with call-back samples.

For a better understanding of probability samples with quotas, however, it is also useful to consider the probability of completing an interview on the first call, as well as the average number of calls required. There is a very substantial increase in the probability of finding a respondent after the first call, since the interviewer will often learn on the first call when the respondent will be available. Thus, the use of average calls required overestimates the probability of a respondent being available on a probability sample with quotas. Table 9.2 shows these first-call probabilities for the surveys of Table 9.1. Naturally, these results are somewhat more variable since they utilize only a fraction of the data, but they show exactly the same relationships.

Table 9.1
Average Calls Required to Complete an Interview on Various Probability with Call-Back Samples[a]

Sample	All respondents	Males	Under 30	30+	Females	Employed	Unemployed
NORC							
All places	2.7 (906)	3.0 (387)	3.2 (76)	2.9 (311)	2.5 (519)	3.0 (212)	2.2 (307)
10 largest SMAs	3.2 (206)	3.2 (96)	3.4 (27)	3.2 (69)	3.3 (110)	3.9 (48)	2.8 (62)
Other SMAs	2.9 (357)	3.3 (155)	3.5 (30)	3.2 (125)	2.5 (202)	2.9 (91)	2.2 (111)
Non-metro counties	2.3 (343)	2.4 (136)	2.8 (19)	2.4 (117)	2.1 (207)	2.6 (73)	1.9 (134)
Survey Research Center							
All places	2.2 (7528)	2.3 (3658)			2.1 (4031)		
Large metro	2.5 (2299)						
Other urban	2.1 (3717)						
Rural	2.8 (1512)						
Britain	2.3 (1443)	2.4 (938)			2.0 (505)	2.3 (55)	2.0 (450)
Elmira	1.9 (1029)	2.1 (452)			1.7 (577)		
Madison	2.0 (743)	2.2 (313)			1.8 (430)		

[a]From Sudman (79).

Table 9.2
Probability of Completing Interview on First Call, by Age, Sex, and Employment Status with Call-Back Samples[a]

Sample	All respondents	Males	Under 30	30+	Females	Employed	Unemployed
NORC							
All places	.28 (906)	.23 (387)	.24 (76)	.22 (311)	.31 (519)	.19 (212)	.40 (307)
10 largest SMAs	.19 (206)	.18 (96)	.26 (27)	.14 (69)	.20 (110)	.10 (48)	.27 (62)
Other SMAs	.26 (357)	.21 (155)	.30 (30)	.18 (125)	.30 (202)	.16 (91)	.41 (111)
Non-metro counties	.35 (343)	.28 (136)	— (19)	.31 (117)	.40 (207)	.30 (73)	.45 (134)
Survey Research Center							
All places	.32 (2963)	.26 (1340)			.36 (1623)		
Large metro	.21 (1724)	.15 (323)			.26 (401)		
Other urban	.32 (1501)	.26 (659)			.36 (842)		
Rural	.42 (738)	.37 (358)			.47 (380)		
Britain	.44 (1443)	.40 (938)			.51 (505)	.35 (55)	.53 (450)
Elmira	.38 (1029)	.24 (452)	.18 (108)	.26 (344)	.49 (577)		
Madison	.40 (743)	.27 (313)	.21 (57)	.29 (256)	.49 (430)		

[a]From Sudman (79).

9.4 SAMPLE BIASES DUE TO QUOTA PROCEDURES

A typical quota procedure involves only a few strata. (The sample used at the National Opinion Research Center had quotas for four strata: men under 30, men 30 and over, unemployed women, and employed women.) Sample biases arise if the probabilities of being available for interviewing vary within a stratum. One can certainly make the strata more homogeneous by splitting off additional strata from those that already exist, but this makes the search procedure more costly. If there are too many strata, it becomes cheaper to make call-backs than to continue the search for a respondent with rare characteristics.

Once the strata have been specified, the quotas are determined from the best available sources, usually census data. Example 9.1 illustrates the procedure for establishing quotas for a given segment.

Example 9.1 Establishing Quotas for a City Block

Suppose a block within Census Tract 1157 has been chosen in the city of St. Louis, Missouri, and the designated sample size for that block is eight respondents. The source of data for quota controls will be found in Tables P-1 and P-3 of the 1970 census tract report for St. Louis, PHC(1)-181. (If the segment is not in a metropolitan area, the same data will be found for the location or rural part of the county in other census reports.)

Table 9.3 gives the distribution of the population in the tract by sex, age for men, and employment status for women. The data on age and sex are taken from Table P-1, and that on employment from Table P-3.

The percentages are multiplied by 8 to give the quotas for the block. Since this multiplication yields decimal numbers, a random number table is used to determine the exact quota. Thus, in Table 9.3, for men 18–29, the percentage 9.7 times 8 is .78. The random number selected (67) is less than 78, so the quota for men 18–29 in this block is one respondent. If a random number larger than 78 had been selected, the quota would have been 0. Similarly, for men 30 and over, the random number 98 is larger than the decimal part of 2.48, so the quota is two, not three, respondents.

Table 9.3
Population Distribution and Quota Controls for Tract 1157

Stratum	Number	Percentage	8 × Percentage	Random number	Quota selected
Men 18–29	370	9.7	.78	67	1
Men 30 and over	1181	31.0	2.48	98	2
Women, unemployed	1369	35.9	2.87	80	3
Women, employed	893	23.4	1.87	–	2
Totals	3813	100.0			8

It is possible both theoretically and empirically to estimate sample biases due to differential availability within strata. Typically, these biases will be on the order of 3% to 5%. For example, comparing call-back and quota samples on NORC studies of attitudes toward the cold war, the differences averaged about 6%.

Of 17 demographic comparisons, there were no differences except sex and household size. Naturally, since there is a quota on sex, the probability sample with quotas matches census data. Both of the call-back samples were deficient in males because of noncooperation. That is, since the cooperation rate among men is lower than among women in the ordinary call-back sample, the sample with quotas is superior for this characteristic.

The comparisons of household size suggest that the quota sample is deficient in one- and two-member households. These results suggest that some of the remaining availability bias in the quota sample could be eliminated by imposing a household size control.

9.5 COSTS OF PROBABILITY SAMPLES WITH CALL-BACKS AND WITH QUOTAS

The chief argument made for the old quota samples was that they are cheap. The costs of probability sampling with quotas are still less than the costs of sampling with call-backs, but the differences are much narrower. A substantial portion of the cost differential is due not to field activities but, rather, to other aspects of the study that are unrelated to sampling.

Comparing the total costs of probability call-back and quota studies, call-back sample costs per case are typically three times higher than the costs per case of quota samples; a substantial part of this difference is a result of differences in planning, processing, and analysis between the two surveys. Almost always, the planning and analysis of call-back samples is costlier and takes a larger part of the study's total cost. It seems clear that it is not the sample design that determines the cost of a study but, rather, the cost that determines the sample design.

To be more explicit, where survey results will receive very sophisticated analysis, or when critical decisions will be based on them, it will be worthwhile to pay a substantial cost to achieve high standards of sampling, processing, and control. Thus, the Census Bureau rightly has very high standards on their current population surveys. On the other hand, many exploratory studies do not require such high standards, since the analysis may be more limited and the questionnaire itself may be a major source of error. Here, quota sampling is justified.

On the other hand, the relationship is reciprocal. One reason that analysis costs are higher on call-back samples is that the analysts spend more time waiting for results to become available. Very often, the field data-collection period is extended for several weeks, which delays the processing for an additional time period. Although this waiting time may sometimes be useful in developing codes and modes of analysis, it is frequently wasted.

If one looks only at total field costs, which include both direct and supervisory costs, the ratio of the costs of probability samples with call-backs to probability samples with quotas drops from 3:1 to 2:1.

This comparison can be carried still one step farther. The major difference between the two types of samples is the cost of supervision. Again, it should be noted that not all of the difference in supervisory cost is a result of the sampling method. Some of this difference can be attributed to the greater quality-control checks generally used in the more expensive samples. The additional effort generally made in training interviewers on call-back samples requires more supervisory time. There are, however, some charactertistics of a call-back sample that do generate greater supervisory costs. Typically, the interviewer is told to make three call-backs and then to check with the supervisor for further instructions if an interview has not been completed. The decision process whereby this occurs is quite costly in supervisory time, as are the letters and long-distance calls that accompany the revised instructions. If methods can be found for standardizing the follow-up procedures and eliminating most of the ad hoc decisions now made, a substantial part of field supervisory costs could be eliminated.

Another aspect of call-back sampling which leads to higher supervisory costs is the time period required to complete the interviewing. Since some respondents will be temporarily unavailable at the interviewer's initial call, a return visit at a later time will be necessary. If the respondent is on vacation or in the hospital, it may be several weeks before the interviewing is completed. Proper allowances should be made for this when scheduling, and the flow of completed questionnaires should be watched carefully. Nevertheless, some added cost due to the stretching-out of the time period cannot be avoided with call-back samples.

Because readers of this book are likely to have limited resources, the use of probability sampling with quotas should be considered carefully when a study is being planned. This procedure has enabled many social scientists in the last decade to do studies that otherwise could not have been afforded, or to increase the sample sizes so that additional analyses were possible. Although this is not a method to use for making major policy decisions, it is a form of probability sampling with well-defined characteristics and is clearly far superior to loose judgment or convenience samples.

9.6 SCREENING FOR RARE POPULATIONS

Although general population samples are still of great importance, there has been an increasing trend toward studies of special populations. This introduces new sampling problems as the populations become increasingly rare, since no lists are available for the special populations. Locating these populations then requires a door-to-door search for a probability sample of the general population. This search is usually called "screening."

Here, a very rare population is defined as one consisting of about 1% or less of the total population. This is obviously an arbitrary cutoff, but it does have some operational significance, as will be shown.

Some examples of very rare populations are:

a. Low-income white intact families with a male head aged 25 or younger in the labor force, and with the head or wife having living parents
b. Black males aged 30–39 in intact families and in the labor force
c. Men or women over age 65 with severe undetected hearing losses

These actual examples from recent surveys suggest the nature of the sampling and screening problems. No lists of these populations are generally available, and census figures are either not available at all, or available only at a state or national level.

Useful discussions of sampling of rare populations have been given by Hansen, Hurwitz, and Madow (35) and particularly by Kish (40), but there has been no detailed discussion of very rare samples. Obviously, an efficient procedure would be to use a very large multipurpose sample, but this will not often be possible. Thus, a general sample of 20,000 cases would be required to yield a sample of 200 cases for a very rare population. If such a sample is available, it should be used, even if the results are a year or two out of date. It would certainly be less costly to trace households who have moved than to screen for a new sample.

Ultimately, population mobility makes it too difficult to use sampling frames that are more than a few years old. Alternatively, it may be efficient to screen a very large sample for a rare population and use the screening information for selecting future samples of other rare populations. The uncertainty about the need for these future samples generally requires that the full cost of the screening be borne by the current study.

Assuming no list or master sample is available, the chief method of reducing the costs of sampling rare populations is to select very large samples within clusters. Since these large clusters yield only small clusters of the rare population elements, and since travel costs are reduced drastically, this is an effective procedure for special populations that are not too rare (say, 2% or more of the total population). This procedure may become very inefficient, however, for very rare populations. A characteristic of such populations is that a majority of the clusters will have either no members of the rare population or only a very small number. Thus, most of the screening effort will be wasted.

Two examples of distributions of very rare populations are given in Table 9.4. Part A of the table gives the 1960 distribution, by census tract, of individuals born in Mexico, or those with Mexican parentage, living in Chicago. Part B gives

Table 9.4
Distribution of Census Tracts in Chicago
for Two Very Rare Populations, 1960[a]

A. Of Mexican parentage		B. Nonwhite divorced males	
Percent population in tract	Percent of tracts	Percent population in tract	Percent of tracts
0	17.3	0	40.9
.01– .10	8.8	.01– .50	20.1
.11– .20	11.8	.51–1.50	6.6
.21– .30	7.4	1.51–2.50	6.6
.31– .40	8.5	2.51–3.50	8.1
.41– .60	9.2	3.51–4.50	7.3
.61– .80	8.8	4.51–5.50	6.9
.81–1.20	7.7	5.51– +	3.5
1.21–1.50	3.7		
1.51–2.50	9.6	Total	100.0
2.51–3.50	3.7	Mean percentage	1.4
3.51– +	3.5	Number of tracts	259
Total	100.0		
Mean percentage	0.7		
Number of tracts	272		

[a]From Sudman (77).

the distribution of nonwhite divorced males in Chicago. The mean percentages of these groups in the total population are 0.7% and 1.4%, respectively. These groups were selected because census information is available, but such information usually is not available, as in the examples cited earlier. Some preliminary screening, followed by disproportionate sampling of clusters, can greatly increase the efficiency of locating very rare population elements. The ideas of Wald (98) on sequential sampling appear to be useful in this preliminary screening and are discussed later.

A special aspect of disproportionate sampling for rare populations is that clusters in which the proportion p of the rare population is less than some cutoff value P_c may be omitted entirely from the sample. This can be justified, even by non-Bayesian samplers, by estimating the maximum bias resulting from such procedure, which is usually small. The Bayesian approach discussed in Chapter 6 indicates that, under the weakest of priors, an optimum sampling design omits clusters in which $p < P_c$ because of cost reasons.

Finally, a consideration of field procedures for screening indicates that classification errors may be expected, and methods for minimizing these errors are suggested.

9.7 OPTIMUM SAMPLING DESIGNS WHEN SCREENING COSTS ARE LARGE

For sampling from very rare populations, one would stratify the sample by the difficulty and cost of locating respondents. Generally, the σ_i's and V_i's will be equal over all strata, so the sample rates depend only on the costs and, thus, on the proportion of the rare population in the stratum. The only way in which this procedure differs from Neyman optimum allocation is that it is likely that some strata may not be sampled.

To illustrate the procedure, we give two examples. In these examples, tracts with zero proportion rare population have been put into the same stratum with tracts having a very low proportion. This gives them a possible nonzero probability of selection because members of the rare population may have moved into the tract since the last census.

Although the procedures do not depend on the specific estimators used, it may be helpful to think of a study of political attitudes, the goal of which is to estimate the proportion of members of the rare population who intend to vote for a given candidate. Thus, all the σ_i's for both examples are chosen = .25 to illustrate a variance of a dichotomous question with a 50–50 split. The V_i also assume equal prior information in each stratum, equivalent to a prior sample of five cases per stratum. A non-Bayesian could interpret these results merely as those of an earlier pilot test with $V_i = \sigma_{\bar{x}}$, the standard error of the mean for a sample size $n = 5$. The costs per interview in each stratum are derived by using estimates of \$2 per screening call and \$10 per interview. These screening costs may seem high, but they assume a 10–15-minute screening interview, and call-backs if no one in the household is available. As shown later, cheaper screening procedures are possible but may lead to serious omissions.

Using Formula (6.4) and assuming a total field budget of \$20,000 leads to the following sample allocations for Mexican-Americans given in Part A of Table 9.5.

In Part A, no strata are omitted, but Stratum a has a very small sample size, $n_a{}^o = 3$. The optimum allocation procedure is approximately that of Neyman optimum allocation, in which the sampling rate is inversely proportional to $\sqrt{c_i}$. Note that, since c_i is very nearly inversely proportional to p_i—the proportion in the total population—this implies that the sampling rate is roughly proportional to $\sqrt{p_i}$. In Part B, Stratum a is omitted entirely, but, in the remaining strata, sampling rates are roughly proportional to $\sqrt{p_i}$.

The variance for optimum allocation is $\sigma^2{}_{\text{opt}\cdot} = \Sigma\,(\pi_i{}^2\sigma^2/n_i{}^o)$, ignoring finite correction terms that are unimportant here. The variance for simple random sampling is $\sigma^2{}_{\text{srs}} = \sigma^2/n$. Since, for both examples, we assume $\sigma_i{}^2 = \sigma^2$, the ratio $\sigma^2{}_{\text{opt.}}/\sigma^2{}_{\text{srs}} = n(\Sigma(\pi_i{}^2/n_i{}^o))$.

Given a budget of \$20,000, one could select a simple random sample of 76

Table 9.5
Sample Allocations for Individuals Born in Mexico or with Mexican Parentage and for Nonwhite Divorced Males[a]

Stratum	Proportion in total population p_i	Proportion of all clusters in stratum	n_i	σ_i^2	V_i	c_i	B_i	C_r	n_i^0	Total screening required
			A. Mexican-Americans							
a	0 – .60	.63	.14	.25	.05	$1132	2404	8970	3	1683
b	.61 – 1.50	.20	.24	.25	.05	222	621	–	27	2862
c	1.51 – 2.50	.10	.26	.25	.05	110	403	–	45	2250
d	2.51 and over	.07	.36	.25	.05	60	215	–	88	2200
		1.00	1.00						163	8995
			B. Nonwhite divorced males							
a	0 – .50	.61	.04	.25	.05	2510	1252	22,615	0	0
b	.51 – 2.50	.13	.15	.25	.05	143	797	634	30	1995
c	2.51 – 4.50	.16	.40	.25	.05	67	205	–	131	3734
d	4.51 and over	.10	.41	.25	.05	39	152	–	179	2596
			1.00							8325

[a] From Sudman (77).

cases in Part A and 122 cases in Part B. Thus, for Part A, $\sigma^2_{\text{opt.}}/\sigma^2_{\text{srs}} = .67$; and, for Part B, $\sigma^2_{\text{opt.}}/\sigma^2_{\text{srs}} = .43$.

These computations assume an initial sample of 5, or its Bayesian equivalent, in each stratum for optimum Bayesian sampling, and of 20 for simple random sampling.

Another way of comparing procedures is to examine the costs of obtaining the same sampling error. For Part B, the cost of simple random sampling required to reduce the sampling error to the level of that for optimum sampling would be $50,000, 2 1/2 times more. For Part A, the cost of simple random sampling would be about $32,000, or about 62% more.

For the discussion in the next section, it is useful to consider the effect on sampling errors of a simple sampling scheme that omits Stratum A entirely and samples the remaining strata at the same rate. For Part A, the reduction in variance over proportional sampling is 30%, or the increase over optimum sampling is about 5%. For Part B, the reduction in variance over proportional sampling is 56%, and the increase over optimum sampling is about 3%.

9.8 ESTIMATING THE PERCENTAGE OF THE VERY RARE POPULATION IN A CLUSTER

Typically, census data will not be available, and one will not know in advance what percentage of the very rare population is in a cluster. Census data for small areas, particularly recent data, can be used to identify some clusters that have zero or very few members of the rare population, and these would not require further screening. Thus, in the examples given earlier, if one were looking for black males, aged 30–39, in intact families and in the labor force, and 1970 census tract information were available, one could omit the all-white census tracts as well as those with no housing units. This does not remove the need for additional screening, however, for determining the percentage of the very rare population in the cluster. For this screening to be efficient, one should be able to identify the clusters with zero or few members of the population as quickly as possible; for this purpose, sequential sampling is suggested.

Screening for a very rare population in a cluster is similar to inspecting for defectives in a product by sampling lots. Wald (98, Sec. 5.3) showed that sequential procedures generally required average sample sizes only half as large as those required by fixed sampling schemes for fixed error levels. The test statistic used is the sequential probability ratio test, which is particularly easy in the case of a binomial distribution.

In the usual inspection situation, particularly where destructive testing is required, the probability of a type I error (α) is set at a low level (.05 or lower) while the probability of a type II error (β) is generally given a higher value. In our case, the reverse is more nearly correct. The α error of misclassifying a cluster having very few members as having many members of the rare population merely leads to additional screening, with some increase in costs. The β error leads to biased sample results and may be more serious.

The procedure may be carried out graphically by drawing two parallel lines defined by a slope s and intercepts h_0 and h_1. The region below the lower line is the acceptance region, the region above the upper line is the rejection region, and the region between lines requires additional sampling. The minimum n for acceptance is h_0/s.

Table 9.6 presents values of the sequential test parameters s, h_0, and h_1 for values of α = .10 and .40, and for various β values, for the two very rare populations previously discussed. An examination of the operating characteristics of these procedures is given in Tables 9.7 and 9.8. Table 9.7 gives the percentage of eligible respondents who are missed if screening is terminated after

Table 9.6
Values of h_0, h_1 for Combinations of α and β in Sequential Tests[a]

α	β	h_0	h_1	Minimum n (h_0/s)
A. Population with Mexican parentage[b]				
.10	.01	2.49	1.27	356
	.05	1.60	1.25	229
	.10	1.22	1.22	175
	.20	.83	1.15	119
	.30	.61	1.08	87
.40	.01	2.27	.50	325
	.05	1.38	.48	197
	.10	.99	.45	142
	.20	.61	.38	87
	.30	.38	.31	54
B. Nonwhite divorced males[c]				
.10	.001	2.42	.82	178
	.01	1.60	.82	118
	.04	1.35	.81	100
	.05	1.03	.80	76
	.10	.78	.78	58
.40	.001	2.28	.33	168
	.01	1.46	.32	107
	.02	1.21	.32	89
	.05	.88	.31	65
	.10	.64	.29	47

[a] From Sudman (77).
[b] p_0 = .0025, p_i = .015, s = .006989.
[c] p_0 = .0025, p_i = .04, s = .01363.

Table 9.7

Rate of Omission and Reduction in Sampling Required If Screening Is Terminated after n Calls If No Eligible Respondent Is Found[a]

n	Percent reduction in sampling	Percent omitted of total eligible	Percent omitted by percent in cluster					
			.01–.50	.51–1.50	1.51–2.50	2.51–3.50	3.51–4.50	4.51+
			A. Nonwhite divorced males[b]					
200	27.6	1.6	.40	.08	0	0	0	0
150	35.7	2.5	.49	.20	0	0	0	0
100	44.6	4.6	.53	.30	.06	.05	0	0
75	51.1	8.5	.71	.44	.11	.05	.11	0
50	61.1	23.5	.73	.76	.31	.38	.21	.07
			B. Population with Mexican parentage[b]					
200	30.6	6.8	.86	.67	.36	.20	.04	0
150	40.4	9.7	.88	.72	.48	.29	.08	0
115	47.5	13.2	.90	.72	.60	.36	.15	0
80	56.9	16.5	.95	.79	.76	.44	.17	0

[a]From Sudman (77).
[b]Clusters of 400.

Table 9.8
Additional Rate of Omission and Reduction in Sampling Required If One But Not Two Eligible Found in n Screenings[a]

Stopping rule[b] (n)	Percent reduction in sampling	Percent omitted
	A. Nonwhite divorced males	
75 – 150	3.1	2.0
100 – 175	3.7	3.7
150 – 220	2.4	2.6
	B. Of Mexican Parentage	
00 – 210	4.5	3.5
115 – 265	3.0	1.3
150 – 300	2.2	1.2

[a]From Sudman (77).
[b]One but not two eligible found in n screenings.

n calls and *no* eligible respondent is found. Table 9.8 shows the omission rate if *one* but not *two* eligible respondents are found in n screenings. It may be seen from Table 9.7 that this part of the sequential procedure has only small effects on the percentage omitted and on the reduction in sampling. In our examples, the effects of sequential procedures if two or more eligible respondents were found were virtually zero.

Tables 9.7 and 9.8 suggest that, for this application, the sequential procedure has almost the same characteristics as those of a double-sampling procedure that merely specifies that screening stops if no eligible respondent is found after n calls.

It should be clear, now, why this procedure is most useful for very rare populations. For other populations in which the proportion of members of the rare population to the total population is .03–.04 or higher, there will probably be very few, if any, zero clusters, and the cost savings resulting from reduced screening may be insufficient to repay for the uncertainties introduced, for the special instructions to interviewers, and for the need to weight the final results.

The results of Tables 9.7 and 9.8 do not represent actual field screening but reflect the results of simulated screening from the distributions shown in Table 9.4. The assumption made in these simulations is that eligible respondents are randomly distributed throughout the cluster. In one case in which there was actual field screening for low-income, white male heads aged 25 or under, there was no evidence of lack of randomness. If very rare individuals are not distributed randomly throughout the cluster, this will increase either the number of screening calls required or the probability of misclassification.

One may raise the more general question of how well the assumptions underlying sequential analysis are met by distributions of rare populations. The basic assumption is that the distribution consists of two binomial distributions with different values p_0 and p_1. One way of looking at this is to examine the operating characteristic curves for sequential tests to those for the screening procedures discussed here. At least for these examples, the OC curves are sufficiently similar so that one would not be concerned about the assumptions.

9.9 FIELD PROCEDURES FOR SCREENING

Since screening is so expensive, there is a natural tendency to attempt to shorten and simplify the screening interview. This can be done by asking a series of screening questions so constructed that a "no" would end the interview. Thus, for the first example of a rare population, the interviewer would first ascertain race, then ask if the household consisted of a husband and wife, next ask if the male head is aged 25 or under, and if he is in the labor force, and finally ask if the head and wife have living parents. The major drawback to this approach is that the respondent quickly recognizes that a very specific type of respondent is sought. It is simple for the respondent to act as a "gatekeeper" and to prevent an interview from being conducted merely by saying "no" to one of the questions. That is, the household is deliberately misclassified by the respondent on the screening interview in order to avoid the perceived bother of being interviewed. Often, the ultimate respondent will be quite willing to be interviewed if one can get past the gatekeeper.

The solution to this problem requires a longer screening interview but prevents the gatekeeper from functioning. The interviewer, after introducing the purpose of the study but not specifying who in the household is to be interviewed, obtains a detailed listing of household characteristics. Thus, for the example just given, the interviewer would

1. List all household members.
2. Obtain the ages of all household members.
3. Obtain the labor force status of all adults.
4. Determine whether or not the parents of the adults in the household are living.
5. Obtain an estimate of household income.

Sometimes, the interviewer is not told what the desired characteristics are, so that neither the respondent nor the interviewer can act as gatekeepers.

Unfortunately, even if gatekeeper errors are avoided, the problem of misclassification is serious. Misclassification errors are not distributed randomly but tend to systematically miss members of the rare population. There are two reasons for this:

1. The final interview is usually sufficiently detailed to ultimately exclude respondents incorrectly classified as being in the rare population on the intitial screening. Those excluded incorrectly on the screening, however, are not available for subsequent interviewing.
2. Classification errors tend to reduce the variability of a population. When the respondent is doubtful, his answer will move toward the perceived mean and away from the rare population.

The magnitude of misclassification errors may easily be underestimated. Previous studies have shown misclassification errors of 5–10% on many variables viewed singly. These errors multiply, however, when the rare population is defined by many characteristics. Thus, for some complex populations, as many as one-third to one-half of eligible respondents may be missed. These errors can be reduced by making the screening questions more detailed, but this adds to the length and cost of the screening interview. Alternatively, one can loosen the criteria for eligibility on the final interview, again adding to the cost of interviewing. Although misclassification cannot be avoided, it can be minimized by careful, thorough screening. The more complete the screening, the greater the likelihood that future studies or rare populations can use the screening sample.

9.10 SNOWBALL SAMPLING

The term "snowball sampling" is used to apply to a variety of procedures in which initial respondents are selected by the probability methods described in the earlier chapters, but in which additional respondents are then obtained from information provided by initial respondents.

Snowball samples have been used in the following social science applications.

Sociometric Studies. The focus is on determining the communication or friendship networks, typically in small groups.

Studies of Elites. Initial respondents are selected by their formal roles, but informal members of the elite are found through snowballing.

Control Groups of Pseudoexperiments. In attempting to determine the effects of participation in some experimental program, such as the Job Corps or a Manpower Training Program, no perfect control group can usually be found because the participants are self-selected. Asking each sample participant for the name of a person like himself or herself who is not in the program yields a control group sample that is very similar to the experimental group for most variables.

Locating Rare Populations. Once some members of a rare population have been located, either through screening or by some special list, others are located through snowball sampling.

SAMPLE BIASES IN SNOWBALL SAMPLING

The major sample bias that results from snowball sampling is that the person who is known to more people has a higher probability of being mentioned than does the isolate, the person known to only a few others. This bias is least important in studies of elites or small groups for which the snowball procedure is continued until no new names are mentioned. In these cases, the probability of missing elements of the selected population is small. Note that, in this case, there is also no sampling variance, because the total universe has been surveyed, unless one considers the group as some subsample from a super population.

When snowball sampling is used to generate control groups for pseudoexperiments, the differences between the experimental and control groups are smaller than when an independent control group is selected by using geographic and demographic variables. Obviously, this procedure should not be used if it is possible to compare accepted applicants to those not admitted into the program simply because of lack of space.

The least satisfactory use of snowball sampling is the locating of rare populations. Here, there may be major differences between those who are widely known by others and those who are not. The procedures given in the previous section seem preferable in this case.

SAMPLING VARIANCES OF SNOWBALL SAMPLES

If it is appropriate to compute variances for snowball samples, the procedures of the last chapter may be used. The lack of independence between sample members is merely another example of cluster effects; and presents no new problems.

If snowball samples are used as control groups, the appropriate statistics are the differences between the experimental respondent and the person named as control. These differences are summed over PSU and the balanced replication procedures are used as before.

For special rare populations that are not sampled completely, snowball sampling variance estimates do not include estimates of possible sample biases, but they do indicate what would be expected from repeated snowball samples selected in the same way.

9.11 PANEL SAMPLING

A panel is a group of respondents who provide information for more than one period in time. The simplest panels involve respondents, as in voting studies, who report before and after an event, such as an election. More complex panels involve respondents who report weekly on a continuous basis on household expenditures or other behavior. The chief advantages of panels are that the sampling variances of measures of change are much smaller for panels than for groups of independent samples and that it is possible to measure changes in individual as opposed to group behavior over time.

The best known social science examples of panels are the voting studies by the Survey Research Center (11), and the studies of educational and career choice and career patterns by the National Opinion Research Center (17) and Project Talent (26).

This section will discuss the special sample design problems that face the researcher who wishes to establish and maintain a panel, particularly a long-term mail panel. To a large extent, the sampling methods used for panels are similar to those used for one-time surveys. There are, however, two critical areas in panel sample design that were not covered earlier. These are:

1. Sample biases due to noncooperation of respondents
2. Maintenance of a panel through time

In addition, clustering strategies for panels may differ from those used for one-time surveys.

9.12 SAMPLE COOPERATION IN CONSUMER PANELS

No survey achieves full cooperation from selected respondents. Governmental agencies generally get more than 90% cooperation. Other survey organizations with well-trained field staffs get between 75% and 90% cooperation depending on the nature of the survey. For periodic, repeated interviews, an additional 5% or 10% loss should be expected from the remaining sample on each subsequent interview. Consumer panels that require the household to keep a continuing record generally have initial cooperation rates of about 60% and continuing cooperation rates of less than 50%. While the possibilities of sample biases are present in one-time surveys, it is clear that these biases are potentially more serious if the cooperation rate is 50% than if the rate is 80%.

The cooperation achieved is, of course, not independent of the recruiting methods used and the tasks required of panel members. It is, however, striking to note that continuing cooperation in several United States panels and in foreign panels also clusters near the 50% rate. When greater efforts are made to get initial cooperation from respondents, there seems to follow a higher dropout rate on a continuing basis.

The question then arises as to the value of the probability selection methods when only half of the respondents will cooperate on a continuing basis. Why not use looser methods and reduce cost? There are very good reasons for using area methods for selection, even with less than perfect cooperation rates. Location is an important indication of a whole complex of facts about the respondent. Where he lives not only gives some indication of his socioeconomic condition, but also indicates the range of decisions available to him.

9.13 SAMPLE BIASES IN PANELS

General information on panel sample biases is not available, but there is data on the characteristics of households whose members do and do not join expenditure panels. Panel cooperation appears to be best in households with more than two members, in households having housewives in the younger age groups, and in households with more education. Comparisons were made in 1959 between cooperators and noncooperators in a national consumer panel. Previous comparisons made in 1950 and published by the Department of Agriculture (84) agree very closely with these results. These comparisons are shown in Tables 9.9 and 9.10.

The 1959 study was based on initial classifying interviews that were obtained before households were invited to join the panel. Consequently, these comparisons do not show the effects of dropout families. (The problems of panel maintenance will be discussed later.) It is clear that the characteristics shown are not independent but are interrelated. Nevertheless, it seems useful to discuss them individually.

Household Size. Panels have difficulty in recruiting small households. The greatest difficulty is with one-member households. Single individuals are less likely to be found at home, and generally have much less interest in such things as food purchasing than do housewives in larger households. Although the work of keeping continuing records is greater for a housewife in a large household, she is less likely to be working outside the house and may find the recordkeeping a pleasant change from the usual routine. If the housewife is unwilling to keep a continuous record, the probability that someone in the household will keep it increases as the number of adult household members increase.

Age of Housewife. Housewives over 55 years old are less likely to join a panel than are young or middle-aged housewives. This same relationship is seen in

Table 9.9
Comparisons of Characteristics of Households Which Do and Do Not Join Consumer Panels (1959 Study)

Sample sizes, 1959	Percentage of panel cooperators (n=4570)	Percentage of noncooperators (n=4588)
A. Household size		
1 member	5.4	11.5
2 members	24.1	29.8
3 members	20.5	19.5
4 members	22.8	19.0
5 members	14.5	10.7
6 members	7.8	4.9
7 members	2.9	2.4
8 members	1.0	1.4
9 or more members	1.0	0.8
Totals	100.0	100.0
B. Children under 6		
1 child	18.4	13.9
2 children	12.5	8.3
3 children	3.8	2.4
4 or more children	1.3	0.9
0	64.0	74.5
Totals	100.0	100.0
C. Age of housewife		
Nearer 20 years old	8.9	8.4
Nearer 30 years old	29.0	22.9
Nearer 40 years old	28.6	24.8
Nearer 50 years old	19.2	20.6
60 years or older	14.3	23.3
Totals	100.0	100.0
D. Own or rent		
Owners	77.8	74.0
Renters	22.2	26.0
Totals	100.0	100.0
E. Type of dwelling unit		
One-family	82.1	76.9
Two-family	8.3	9.2
Three-family or apartment	7.3	12.0
Other	2.3	1.9
Totals	100.0	100.0

(continued on next page)

Table 9.9 (continued)

Sample sizes, 1959	Percentage of panel cooperators (n=4570)	Percentage of noncooperators (n=4588)
F. Housewife employment		
Employed	30.5	35.7
Unemployed	69.5	64.3
Totals	100.0	100.0
H. Mailed in boxtops or coupons		
Mailed in boxtop or coupon in last 3 months	19.3	13.2
Did not	80.7	86.8
Totals	100.0	100.0
I. Price consciousness		
More price conscious than average	41.2	32.8
Equal	35.0	34.8
Less price conscious than average	16.1	21.4
No opinion	7.7	11.0
Totals	100.0	100.0

one-time surveys in which older people have a higher refusal rate. This also may be related to education and the ability to keep written records. Panel maintenance experience indicates, however, that older households once recruited are less likely to drop out of a panel. The housewives in the 25–34 age group who are most likely to join a panel are also the most likely to drop out as a result of changing family circumstances, particularly a new baby.

Own versus Rent. Panel cooperators are slightly more likely than noncooperators to own their own homes. This is due mainly to differences in household size.

Type of Dwelling Unit. Again due to differences in household size, panel cooperators are more likely than noncooperators to live in single-family dwelling units.

Housewife Employment. Employed housewives are somewhat less likely than nonemployed housewives to have the time or interest to join an expenditure panel. These employed housewives are also more likely to come from small households.

Table 9.10
Comparison of Characteristics of Panel Households with Characteristics of All U.S. Households Based on 1960 Census and Current Population Surveys

	Percentage of panel households	Percentage of all U.S. households (based on Census Bureau)
A. Household income, 1961		
To $1,000	2.5	5.0
$1,000–1,999	6.6	8.0
$2,000–2,999	7.6	8.7
$3,000–3,999	8.9	9.8
$4,000–4,999	11.5	10.5
$5,000–5,999	13.8	12.9
$6,000–6,999	13.1	10.8
$7,000–7,999	10.4	8.7
$8,000–9,999	12.0	11.3
$10,000 and over	13.6	14.3
Totals	100.0	100.0
B. Occupation of household head		
Laborers except farms and mines	2.3	4.4
Operatives and kindred	16.8	16.0
Professional, technical, and kindred	11.2	9.4
Managers, officials, and proprietors	9.5	12.5
Clerical and kindred	9.3	6.2
Sales workers	5.4	4.7
Servants and private household workers	5.5	6.1
Craftsmen and foremen	18.2	16.0
Farmers	5.9	6.5
Retired, in school, or unemployed	15.9	18.2
Totals	100.0	100.0
C. Education of household head		
0–8 years completed	30.3	37.5
9–12 years completed	46.5	44.3
13 or more years completed	23.2	18.2
Totals	100.0	100.0
D. Type of household head		
Male head, spouse present	85.8	87.4
Male head, no spouse present	0.7	2.6
Female head, no spouse present	13.5	10.0
Totals	100.0	100.0

Save Trading Stamps. A higher proportion of panel households save trading stamps than do panel noncooperators. The saving of stamps is related both to greater price consciousness and to greater organization of activities by the housewife.

Mail in Boxtops or Coupons. The mailing of boxtops or coupons to receive premiums shows a greater interest in home management as well as some price consciousness. The process of filling in a form and mailing it is related to keeping a written record as a panel member. It is, therefore, not surprising to see in this table that panel housewives are more likely to mail in boxtops or coupons than are noncooperators.

Price Consciousness. A satisfactory measure of price consciousness requires observation of the actual purchase behavior of shoppers. Since this was not available, the respondent was asked for her self-evaluation as to whether she was more price conscious than average. More panel housewives considered themselves price conscious than did panel noncooperators.

Household Income. There is a good agreement, seen in Table 9.10, between the income distribution of panel households and all U.S. households as estimated in the Current Population Report P-60 series. Only at the very lowest end of the income distribution, among households with incomes of less than $1000, is there much difference.

Occupation of Household Head. There do not appear to be major differences in the occupations of heads of consumer panel households as compared to all U.S. household heads, again as measured in the Current Population Report P-60 series.

Education of Household Head. As one would suspect, panel household heads have had more education than heads in the general U.S. population. Education within the household is clearly related to the ability and interest in keeping written records. It is interesting to note, however, that the differences between the panel and the general public on this characteristic have been steadily decreasing as older, less-educated people die, and as the general level of education in the United States rises.

Type of Household Head. This table shows that panels underrepresent households composed only of a single male. This difference would be more serious if these households were not such a small part of the total population.

Except for household size, most panel biases are small for demographic characteristics. The greatest differences between panel cooperators and noncooperators are seen in social–psychological characteristics, such as degree of organization and price consciousness.

9.14 PANEL MAINTENANCE POLICIES

The basic point to remember in maintaining a long-term continuing sample is that such a panel is dynamic and is not to be treated as a sample drawn for a one-time survey. To repeat the obvious, universe characteristics, in total, change very slightly from year to year but the universe itself is changing continuously as a result of new household formations, dissolution of old households, and household moves. A panel, if it is to remain representative of this changing universe, must reflect these changes. A more positive statement can be made, however: A panel, if it is representative of a universe at a fixed point in time, does reflect these changes and can, thus, not only remain continually representative of the universe but also provide evidence as to what changes are occurring. That is, it can do these things if it is allowed to do so. To illustrate this point, consider:

1. Household moves
2. Household dissolutions
3. New household formations
4. Panel member dropouts

Let us take the first question, concerning what to do with families who move. People who are accustomed to working with a good fixed sample frame become extremely uneasy when this area sample frame is discarded. Yet, in operating a panel, experience has shown that it is essential that families who move be followed when they move. This may, perhaps, be conceived as a dynamic system that is frozen for just an instant to allow a sample to be drawn from it. The system is then released and the motion of the sample represents the motion of the universe. If one made the mistake of sticking to a fixed sample frame, thus dropping the households that moved out of the selected dwelling units and replacing them with households that moved in, the following difficulties would arise:

1. It would be difficult to locate and include dwelling units that were built after the sample frame was designed.
2. It would be difficult to allow for shifts in population from some states to others, notably the population shift to the West.
3. Dropping families who move frequently would require costly or impractical methods to prevent obvious sample bias, certain types of families (i.e., young, small households) being far more likely to move than larger or older families.

Allowing a panel that is representative of the universe at a given point in time to move, that is, not dropping or replacing households when they move, gives one a sample that is geographically correct at all times. There is a question as to whether any original nonrepresentativeness of a panel affects its ability to adequately represent moves in the universe. Fortunately, there seems to be empirical evidence that a well-designed panel, even with some sample selectivity, does a good job of representing moves in the universe, since these moves do not appear to be highly correlated with willingness to join a panel.

For panels that are interviewed periodically, it is important to obtain the name of a close friend or relative who will always know where the respondent lives, so that movers between interviews may be traced. In addition, periodic mailings to respondents not only keep cooperation high but also provide information on new addresses for movers.

If the interviewing must be done face-to-face, a respondent who moves to a new PSU where no trained interviewer is available will be lost from the panel. Frequently, however, it will be possible to get all or most of the required information from such respondents either by long-distance telephone or by mail. In some cases, it may even pay to send an interviewer from the nearest PSU to the new location, even if the travel costs are considerable.

For some studies, a panel may be limited to residents of a specified geographic area, such as a city, county, or state. If a household moves from the area, it becomes ineligible and is dropped from the panel.

Accounting for moves, however, is not enough. Some method must be designed for continually rejuvenating a long-term panel by bringing into it the proper number of new households formed and by dropping from it dissolved households. The dissolved households are quite easy to handle. The only necessity is that a careful watch be kept of all moves a family makes, in order to drop from the panel any families who move into an established household rather than establishing a new household. The situation in which all members of a household die at the same time is rare. Far more typical is the case in which, when the husband dies, the wife goes to live with some other members of her family. If she has been a panel member, she is dropped at this time. There is, of course, no need to replace her in the panel with a new household.

A panel can also provide information of new household formations. Usually, the household reports periodically whether or not there has been any change in the number of adults or children living in the home and if anyone has moved away to set up housekeeping. The reason for moving away is given also. By far the greatest majority of these moves to set up housekeeping are results of marriages. Family members who move away to set up housekeeping are recruited with probabilities inversely proportional to the number of persons who will constitute the new household. (This is done so that all new households have the same probability of being added, regardless of the size of the new household.)

Thus, in the case of a marriage, half the split-offs are recruited. Again, empirical evidence has indicated that the panel adequately reflects the new household formations in the universe—this recruiting method constantly brings new young households into a panel at the proper rate.

There are some changes that occur in panels that do not reflect changes occurring in the universe. The first of these is that panel members occasionally drop from the panel after they have been recruited, for reasons beyond the control even of the best-run panel. Personal situations like illness in the family or the birth of a child are some of the more common reasons for dropping. Although this dropout rate is less than 1% per month, or 10% per year, it may have an effect on panel data unless methods for replacing these households are used. Two methods of handling this situation are available:

1. One might oversample on the basis of past experience with dropout rates and their connection with various family characteristics. In practice, it is not economical to maintain a large oversample whose data are not used, so most panels replace households when they drop out.
2. In the replacement of households, the same kinds of problems are found that are found in originally selected households—that is, noncooperation. Basically, the method specifies a directed route that is followed until a cooperating household is found with characteristics similar to that of a dropout family.

It should be pointed out that the reduction of the dropouts in a panel to less than 10% per year is not an automatic process but one that requires considerable effort and experience in the techniques of maintaining panel cooperation. Even though most panels find it important to compensate families with money or prizes, a continuing program of communication with panel families is equally essential to establish and maintain the high level of panel morale that reduces panel turnover.

The use of compensation and of careful initial contact and training is especially important for less-educated respondents for whom keeping written records is a more difficult task. Most of these families, once they are past the initial learning of their task, become extremely loyal panel members.

9.15 PANEL CLUSTERING

How much clustering should there be in a panel sample? There is really no good reason for close clustering if panel households return diaries by mail. If, for any reason, a household resigns from the panel and must be replaced, there is hardly a chance that it would be necessary to replace a nearby household at the same time. The only cost saving would be in the initial travel costs, but due to the lengthy process in recruiting a single household, there is generally very little travel between segments in one day.

On the other hand, the costs of hiring and training interviewers for initial panel recruiting still suggests the same procedures for selecting PSUs and about the same number of PSUs as for other kinds of studies. Within PSUs, however, mail panels should have little or no clustering or respondents. For face-to-face panels, the same clustering as that discussed in Chapter 4 is appropriate.

9.16 SAMPLING VARIANCE COMPUTATIONS FOR PANELS

The procedures described in the previous chapter, using balanced replication of PSUs, are also used for computing variances for panel samples. For measures of change from one period to another, the units of observation are the changes in behavior or attitude at the individual respondent or household level. These change measures are summed over PSU as before. Fortunately, even though there may be substantial sample biases that affect the estimates of behavior or attitudes at a single point in time, the estimates of change have been found to be subject to much smaller biases. Thus, total errors in estimates of change for panels, including both variance and bias, are far smaller than estimates of levels at one point in time.

9.17 SUMMARY

This chapter discussed four special topics that may be useful to social scientists. The first section dealt with certain quota procedures and attempted to demonstrate that they are very similar to traditional probability sampling. Quotas are shown to depend on availability for interviewing and data are presented to show that age, sex, and employment status are reasonable predictors of availability. Quota methods are not unbiased, but the bias is generally on the order of 3% to 5%. The direct cost differences between quota and call-back procedures are small, although indirect savings due to reduced time spent by supervisors may be substantial. The major advantage of this procedure is in exploratory studies and in those studies for which speed of interviewing is a critical factor, as in times of crises or during pre-election studies.

A discussion of screening for rare populations suggested that the usual procedure would be to have a very large multipurpose sample or to screen by using extremely heavy clustering. This becomes very expensive for very rare populations where most of the effort is wasted because no or very few members of the rare population are located. Costs may be reduced substantially if sequential sampling methods are used to locate blank clusters and if these then are eliminated from the sample after only limited screening.

The next section described the uses and limitations of snowball sampling. Snowball sampling is most useful in studies of elites and small group interaction, and is helpful for selecting control groups for experiments in which better procedures are impossible. Because snowball sampling misses isolates, it is not very useful for locating rare populations.

The last section discussed the special sampling problems of panel sampling, primarily sample biases and panel maintenance. For households keeping written records of expenditures, cooperation is higher for younger, better-educated households with more than two members. There is also some evidence that purchase panels get better cooperation from organized, price-conscious households. In maintaining a long-term panel, procedures must be found for tracking household moves, dissolutions, and new household formations. A well-designed panel is capable of tracking these changes in the universe and thus rejuvenating itself.

9.18 ADDITIONAL READING

Perhaps the best-known example of a modified quota sample is the Gallup Poll. A detailed description of Gallup election survey procedures is given by Perry in the *Public Opinion Quarterly* (63). An illustration of the use of quota procedures in Britain is given by Clunies-Ross (13). The discussion of quota sampling here is taken from a more detailed discussion in my earlier book, *Reducing the Cost of Surveys* (79).

The discussion of sampling of very rare populations is taken from a paper in the *Journal of the American Statistical Association* (77). Kish (40, Sec. 11.4) gives a very helpful discussion of sampling for rare, but not very rare, elements.

The initial paper on snowball sampling was by Leo Goodman (33). This is a sophisticated mathematical discussion of the sampling of sociometric networks, and may be too difficult for many readers.

Readers interested in other aspects of panel operations may find two of my papers in the *Journal of Marketing Research* useful (78). There is also a bibliography of other panel studies listed there. Kish (40, Sec. 12.5) has a brief discussion of panels but a fuller discussion of rotating designs with partially overlapping samples. This is the procedure used in the Current Population Survey.

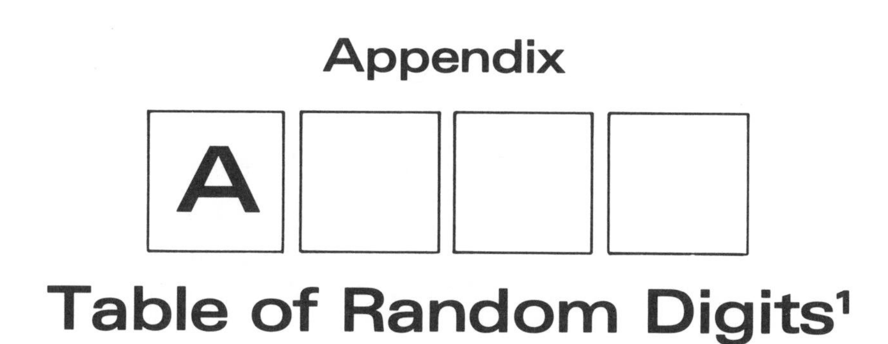

Appendix A
Table of Random Digits[1]

26804	29273	79811	45610	22879	72538	70157	17683	67942	52846
90720	96215	48537	94756	18124	89051	27999	88513	35943	67290
85027	59207	76180	41416	48521	15720	90258	95598	10822	93074
09362	49674	65953	96702	20772	12069	49901	08913	12510	64899
64590	04104	16770	79237	82158	04553	93000	18585	72279	01916
06432	08525	66864	20507	92817	39800	98820	18120	81860	68065
02101	60119	95836	88949	89312	82716	34705	12795	58424	69700
19337	96983	60321	62194	08574	81896	00390	75024	66220	16494
75277	47880	07952	35832	41655	27155	95189	00400	06649	53040
59535	75885	31648	88202	63899	40911	78138	26376	06641	97291
76310	79385	84639	27804	48889	80070	64689	99310	04232	84008
12805	65754	96887	67060	88413	31883	79233	99603	68989	80233
32242	73807	48321	67123	40637	14102	55550	89992	80593	64642
16212	84706	69274	13252	78974	10781	43629	36223	36042	75492
75362	83633	25620	24828	59345	40653	85639	42613	40242	43160
34703	93445	82051	53437	53717	48719	71858	11230	26079	44018
01556	58563	36828	85053	39025	16688	69524	81885	31911	13098
22211	86468	76295	16663	39489	18400	53155	92087	63942	99827
01534	70128	14111	77065	99358	28443	68135	61696	55241	61867
09647	32348	56909	40951	00440	10305	58160	62235	89455	73095
97021	23763	18491	65056	95283	98232	86695	78699	79666	88574
25469	63708	78718	35014	40387	15921	58080	03936	15953	59658
40337	48522	11418	00090	41779	54499	08623	49092	65431	11390
33491	98685	92536	51626	85787	47841	95787	70139	42383	44187
44764	14986	16642	19429	01960	22833	80055	39851	47350	70337

[1] Source: Rand Corporation (67, pp. 145, 189, 230, 371).

96779	94885	33674	52860	39750	47056	59836	10552	26093	40520
06973	61333	00465	70079	02538	83123	86995	05706	71111	40435
22366	71653	64852	69137	36552	25495	85845	71503	31631	58633
37197	91054	45316	64212	63635	68992	02608	93110	21593	56327
15234	35530	10147	65273	07553	78481	62311	36134	89043	56110
75554	64074	37544	34863	36478	79281	58549	44237	19801	31240
47230	79000	08569	74977	06680	99658	07458	17435	08308	11027
30159	83599	72906	07861	13625	35611	03043	69904	55051	74144
28979	73275	87178	48764	58960	40528	14378	03612	90075	96905
65855	05534	44208	08903	19491	82126	66860	32840	54979	22213
95348	50091	44611	49700	54373	80200	76787	16563	18303	66995
41774	64236	05346	57370	74027	46196	05323	43858	84458	81397
03354	96795	86666	35232	38206	24653	39718	80864	28193	86369
88886	09883	77679	07972	20542	81125	54583	70123	13780	74558
48189	54316	64441	32520	06350	71271	93086	52857	63361	98260
29323	88380	34403	29290	29057	74103	18949	37051	93231	73949
57944	15793	46141	77291	54098	37292	71554	16467	07860	47556
26473	35895	03768	48263	09733	22819	43269	63159	38560	13548
90941	14121	32494	52627	65420	12249	66149	47064	51607	98475
15200	48466	68764	30111	29052	75579	92279	88993	69782	27641
03704	21488	23373	27179	78622	98536	85425	92276	97238	28716
06976	19232	77725	26152	82770	07884	32089	25244	20896	06246
58784	61149	89620	88225	38005	81411	29645	40186	35101	89938
92687	63644	39013	63475	45033	98679	44963	28862	51162	71792
68635	28907	63317	16301	35291	27832	49665	26975	36918	71635
25136	53356	21610	96745	14276	83374	38793	27121	02809	18908
10939	52366	77537	80180	98287	14191	09983	42701	69101	73946
98361	61960	02082	44879	33803	64194	41519	20487	22554	69494
34201	75389	40418	63925	01612	60875	27928	54277	23320	23997
94946	95350	19640	24501	58261	86334	12535	12853	97546	80748
92459	46807	00742	98068	05715	91914	30368	76830	01471	31879
01990	61688	21317	58136	81372	32479	89450	54188	15032	52447
56357	03811	04824	53455	88755	30122	02839	71763	49139	06246
36783	05002	71761	35852	40640	62630	26769	02587	44623	95577
88822	11796	28561	27091	93013	64939	94299	98240	57450	18672
03478	89017	30466	54463	32998	45826	92196	84866	90728	60701
15272	84614	27404	33686	51283	72980	53589	61318	78649	06703
29596	47534	89805	95170	89816	58314	03649	64285	14682	12486
71904	81693	94887	45573	76874	74548	36851	48630	77916	78922
05201	51312	78986	27330	63194	98096	93212	74891	55099	02679
16510	95406	39078	31468	43577	67990	11287	27068	37874	61734
83816	94852	73159	76123	05010	08393	62827	13728	34709	39578
19962	86326	99855	14146	28341	93570	34163	59623	14103	63367
66852	52392	32115	75977	80723	96562	19388	64446	73949	83823
84161	37020	79694	35717	73417	15617	93437	46981	94838	12418
58837	30960	84272	38937	27926	95403	61816	32202	11343	99925
12971	62671	87151	80924	08413	22879	51701	84303	65556	20152
21036	13175	77916	31978	78898	69869	22225	13043	49858	81615
34152	24555	54366	40704	33111	00490	53198	52317	77478	38052
50434	17800	99805	32819	71033	83674	84640	67470	60922	25920

74643	91686	64861	13547	47668	02710	11434	82867	40442	23126
30774	56770	07259	58864	02002	78870	29737	79078	03891	96198
52766	31005	71786	78399	41418	73730	44254	81034	81391	60870
30583	57645	02821	46759	21611	81875	75570	71403	95020	90567
11411	87781	95412	14734	68216	24237	64399	57190	62003	08072
65154	65573	06505	85246	28223	48663	84092	80996	62804	25062
71484	49166	54358	28045	90602	26369	18826	34129	11186	02587
36886	15978	25701	88856	99666	72497	28170	74573	66399	98915
31911	32493	55851	22810	77446	47338	58709	00366	76974	89213
57668	83978	67201	95886	02009	87160	63753	12256	84441	23567
20180	80993	05486	83908	29691	75989	16955	24709	66116	55376
29450	78893	24478	40084	96185	64091	74278	19220	59232	79651
10645	25607	05493	66388	14886	10433	13541	60814	84317	56135
86989	65289	55234	46428	57719	18708	88916	98692	40281	81694
81822	31790	27929	60106	04794	50792	52855	69708	54471	98480
57260	73820	40482	50328	08141	63218	92180	33241	88052	99353
03162	15444	20152	57789	87027	24196	69223	03376	28451	60351
03883	01325	75192	63458	69469	82978	39120	56925	58287	37961
20476	36163	49805	39896	40557	89825	99027	68148	68330	14547
04097	82269	13198	82429	30119	06488	40897	77511	82718	20536
03150	81213	38131	72824	38659	60749	64581	64225	07982	13359
41868	08277	15733	03512	66062	55144	42684	92562	95855	18976
19019	75509	82239	46407	80331	67153	97832	07365	78527	25388
66762	65374	73880	42723	52871	61036	35039	70330	19690	65487
56984	25574	51915	47671	32288	94925	46278	62789	66452	20813
62919	32771	60512	67786	62409	25006	15544	27585	41141	07056
80152	53210	58708	63052	29172	61110	50802	61103	31451	73705
14609	03458	36701	61286	55876	76651	12026	57579	63041	00518
69685	96134	85288	20667	85030	25703	24172	75414	94525	98963
37653	41665	23805	16495	50566	92923	58570	37989	14454	96483
97333	40313	38311	79632	88471	94287	18842	56481	10727	78168
46918	35923	43219	89408	42015	70960	39767	33981	00896	62632
77647	92375	95821	56223	20137	50993	01956	37476	65479	37315
84557	54389	92845	66027	25750	75426	74213	54278	87040	87720
50184	53941	77795	47527	99423	48280	94101	96132	65778	57536
62994	33610	61432	81063	07104	72979	67234	51208	58087	64686
74536	61392	54720	01452	23781	37295	24279	36401	84360	97841
24945	28226	08113	79223	78135	43679	40184	67041	13070	51304
76597	61745	07848	59773	88922	31500	89386	01970	31954	74586
21739	73880	84999	71712	20223	04734	05297	38494	57925	83158
76373	36578	07987	73464	86703	43769	38113	85094	76527	89307
75771	17498	31380	65347	86809	22856	80806	83634	08719	34906
37509	95478	66738	53649	66346	55218	73532	43708	97621	39974
26800	77759	78505	67784	64526	05422	54794	98671	74839	79856
65925	13968	71642	98512	87510	56434	93220	76328	53413	17961
50573	71610	48683	29869	32535	75387	22438	84636	94631	27382
83160	02118	84936	15513	18912	48738	72173	28797	17683	88883
80436	58377	66896	58495	27405	17933	89367	75965	93790	58615
74836	84165	82436	68509	32923	12254	28278	69602	49651	70140
80971	88014	58161	70037	08593	81048	90612	70159	47830	03778

55476	47763	58355	18787	42819	72513	01205	47364	75630	55740
43884	76826	19846	64683	47233	40972	74440	47589	80005	01014
31010	42482	01039	45235	23073	55750	55123	79203	54117	01873
80700	22511	03551	43755	87713	55858	68347	22913	01811	06107
79842	31290	03962	29334	11307	10571	33084	78991	31429	28284
25827	93814	42955	90007	74499	09563	36246	27882	63933	26885
56382	86236	11947	65109	47149	22961	95604	05501	90765	92636
09024	52758	06681	32966	35093	64721	54517	26843	83743	10495
31247	73603	84887	70797	62382	64108	74700	26052	05777	73171
02708	02038	66724	49587	56424	12566	54918	93543	73911	85161
71514	81153	13812	42052	94611	80059	54274	48954	92314	52145
44099	91469	85636	54622	45742	08973	24636	63045	81900	57654
20208	38903	24831	25635	50992	51064	14741	29622	75610	07463
10785	63236	84969	71763	19351	08306	23421	55192	09512	22483
25939	19706	61437	26923	29199	10159	99902	73154	60473	38916
78195	11500	93466	00471	77007	88524	03148	82397	32725	53440
15280	87108	99543	10803	07496	48635	09339	06418	43962	32531
80864	47930	36801	09704	12670	88226	89033	74493	24452	26634
53812	63081	82166	18391	62931	07534	48230	91937	77610	85007
94446	05942	23333	36363	68380	94220	35520	17709	16763	84188
34367	83035	88962	85995	07520	39923	42368	93083	08232	13097
43869	75513	08024	27573	87633	32803	50058	43051	95088	49204
84116	30222	14154	98077	71197	98143	75613	66791	62724	35884
89147	86827	01031	45982	74159	49227	67242	14934	50108	73282
65894	90163	78027	39511	40870	39947	06712	94875	56233	19214
67109	11473	10248	70538	69851	92940	78913	62593	81365	54749
30530	20241	61515	72714	61405	55642	72425	36870	22717	67372
22321	86216	64168	59440	58453	13454	38592	76417	06563	93883
31639	53391	72185	05820	60047	78889	30330	53961	67527	63541
38931	51087	58713	02043	91426	24151	98078	18890	75063	26059
91420	44613	89942	43797	76137	90015	75940	22911	55661	68223
67521	49325	28320	48008	23291	62120	31312	05519	68562	47172
72050	53977	45790	77800	31125	83336	43839	76362	93470	85504
72937	85023	88903	63245	05367	98316	78990	07653	37788	76298
14112	75131	00183	87035	21426	39586	64095	62965	97591	92836
39038	10722	23351	04417	28600	68032	27942	79886	25962	94364
46329	44308	97190	79771	21690	80216	82416	81152	69220	92904
94236	45872	61654	96233	22763	98424	81148	14866	99286	08701
58742	09264	73509	01599	14867	81620	55950	41058	98298	51376
16034	43964	73346	52223	57916	38185	58165	96077	10179	22429
95939	03223	12164	17564	50602	53688	49965	67590	72012	97647
67463	03968	04969	92371	79770	92335	34208	45621	32508	41936
77534	21703	98423	23951	54143	33964	20999	54720	29410	08949
74190	30749	01727	92580	69097	40301	69968	47234	76786	68252
72712	34686	63392	04913	50740	02285	52856	71092	31794	10700
25507	87077	33556	22191	28716	94391	79046	10123	37876	00652
21909	11618	19466	68246	52964	04205	03549	79059	26007	36649
75789	08560	32560	56601	88872	95280	20563	70525	81071	22205
77333	62869	95869	55703	94246	30844	36243	93116	93298	85397
94015	96250	92984	89944	43205	27361	01194	65935	27808	34770

Appendix

Measures of Homogeneity ρ

Table A

Measures of Homogeneity ρ between Households[a]

Characteristic	Measure of homogeneity for household clusters of			
	$\bar{n} = 3$	$\bar{n} = 9$	$\bar{n} = 27$	$\bar{n} = 62$
Proportion of households reporting:				
Home owned	.170	.171	.166	.096
Low rent	.235	.169	.107	.062
High rent	.430	.349	.243	.112
Average size of household	.230	.186	.142	.066
Proportion of persons who are:				
Native white males	.100	.088	.077	.058
Males unemployed	.060	.070	.045	.034
Males 25–34 years old	.045	.026	.018	.008

[a]From Hansen, Hurwitz, and Madow (35, Vol. I, p. 264, Table 3). By permission of John Wiley and Sons, Inc. Based on a sample from the selected cities over 100,000 population in the United States.

Table B
Some Illustrative Farm Characteristics with Relatively High Values of ρ[a]

Characteristic	Average size of cluster $\bar{n}$	ρ
Average farm population per cluster in the West	3 hh.	.40
Proportion of farms in the North reporting	3.88 farms	
Woodland		.44
Dairy products		.43
Proportions of farms having commercial orchards in North Carolina	5 farms	.38
Proportion of farms in the South reporting	5.72 farms	
"All other" land		.63
Irish potatoes		.40
Sows		.39
Proportion of farms in the West reporting	4.26 farms	
Barley		.49

[a]From Hansen, Hurwitz, and Madow (35, Vol. I, p. 265, Table 4). By permission of John Wiley and Sons, Inc.

Table C
Measures of Homogeneity between Employee Groups[a]

Characteristic	ρ
Medical charges per family	
For physician services	.024
For hospital services	.01
For all purposes	.015
Percentage wanting increased benefits	.02
Percentage wanting increased benefits and willing to pay more	.005
Number of hospital admissions	.01
Number of surgical procedures	0

[a]From Anderson (3, pp. 139–44). Cluster size = 20.

Table D
Measures of ρ from the National Health Survey[a]

Characteristic	Intraclass correlation ρ		
	$\bar{n} = 6$	$\bar{n} = 9$	$\bar{n} = 18$
Demographic statistics			
Person items			
Total population in households	.270	.228	.196
Negro	.490	.487	.472
All other races	.449	.434	.453
Persons under 1 year of age	.020	.034	.041
Persons under 17 years of age	.222	.164	.137
Persons 17+ years of age	.216	.188	.182
Males 17+ years of age	.082	.084	.095
Total males	.179	.156	.136
Total females	.220	.189	.156
Family income less than $2000	.102	.085	.091
Family income $5000 or more	.236	.211	.180
Household items			
Negro and "other" race head of household	.612	.588	.591
Household income less than $2000	.098	.061	.079
Household income $5000 or more	.238	.221	.200
Labor force			
Current activity, employed	.064	.089	.122
Current activity, unemployed	.029	.019	.040
Health statistics			
Persons with 1+ condition	.165	.122	.122
Males with 1+ condition	.082	.043	.056
Number of chronic conditions for males	.045	.008	.032
Number of chronic conditions for females	.049	.045	.041
Number of bed days in last 2 weeks	.022	.038	.011
Restricted activity days in last 2 weeks	.051	.033	.005
Persons 17+ years who now smoke more than 20 cigarettes per day	.050	.045	.057
Persons 17+ years whose length of time since last regularly smoked cigarettes is 6 months	–.003	–.003	–.009
Persons 17+ who ever smoked 60 cigarettes per day during heaviest smoking	–.015	–.009	–.009
Condition statistics			
Number of acute conditions			
Total persons	.035	.059	.060
Negro persons	.070	.118	.096
Males	–.029	.014	.024

[a]Source: U.S. National Center for Health Statistics (96).

(continued on next page)

Table D (continued)

Characteristic	Intraclass correlation ρ		
	$\bar{n} = 6$	$\bar{n} = 9$	$\bar{n} = 18$
White females	.043	.028	.037
Persons with 1+ bed days	.037	.023	.035
Number of persons injured	−.042	−.017	.000
Number of persons with:			
Tuberculosis	−.066	−.022	−.017
Asthma	.040	.020	.004
Diabetes	.010	.015	.011
Diseases of the heart	−.037	−.037	−.016
Chronic bronchitis	.028	.041	.024
Total conditions (acute plus chronic)	.091	.067	.080
Hospital statistics			
Number of short-stay hospital discharges in past 6 months for:			
Total persons	.001	.022	.032
Persons 45–64 years old	−.003	.027	.059
Persons under 14 years old	−.072	−.030	−.016
Females	−.023	−.0054	.001
Negro and all other races	.095	.089	.086
All persons with family income under $4000	.012	.041	.022
Surgery	.033	.027	.025
Number of short-stay hospital discharges in past 12 months for:			
Total persons	.057	.069	.077
Persons 45–64 years old	.033	.028	.056
Persons under 14 years old	−.044	−.011	−.010
Females	.089	.077	.065
Negro and all other races	.114	.132	.114
All persons with family income under $4000	.039	.065	.060
Surgery	.060	.068	.068

Appendix

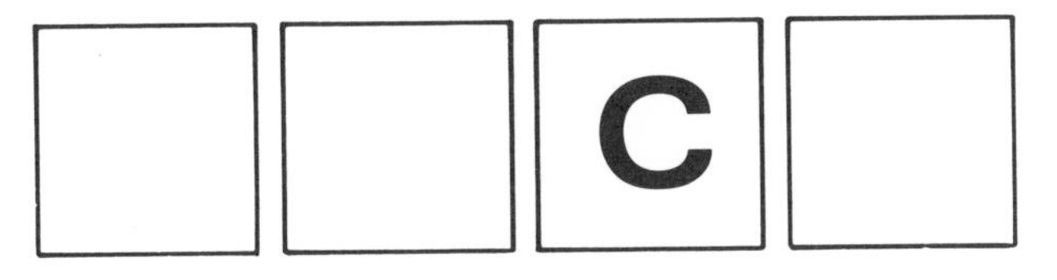

Unit Normal Loss Integral[1]

$$G(u) = P'_N(u) - uP_N(\tilde{u} > u)$$

D	.00	.01	.02	.03	.04	.05	.06	.07	.08	.09
.0	.3989	.3940	.3890	.3841	.3793	.3744	.3697	.3649	.3602	.3556
.1	.3509	.3464	.3418	.3373	.3328	.3284	.3240	.3197	.3154	.3111
.2	.3069	.3027	.2986	.2944	.2904	.2863	.2824	.2784	.2745	.2706
.3	.2668	.2630	.2592	.2555	.2518	.2481	.2445	.2409	.2374	.2339
.4	.2304	.2270	.2236	.2203	.2169	.2137	.2104	.2072	.2040	.2009
.5	.1978	.1947	.1917	.1887	.1857	.1828	.1799	.1771	.1742	.1714
.6	.1687	.1659	.1633	.1606	.1580	.1554	.1528	.1503	.1478	.1453
.7	.1429	.1405	.1381	.1358	.1334	.1312	.1289	.1267	.1245	.1223
.8	.1202	.1181	.1160	.1140	.1120	.1100	.1080	.1061	.1042	.1023
.9	.1004	.09860	.09680	.09503	.09328	.09156	.08986	.08819	.08654	.08491
1.0	.08332	.08174	.08019	.07866	.07716	.07568	.07422	.07279	.07138	.06999
1.1	.06862	.06727	.06595	.06465	.06336	.06210	.06086	.05964	.05844	.05726
1.2	.05610	.05496	.05384	.05274	.05165	.05059	.04954	.04851	.04750	.04650
1.3	.04553	.04457	.04363	.04270	.04179	.04090	.04002	.03916	.03831	.03748
1.4	.03667	.03587	.03508	.03431	.03356	.03281	.03208	.03137	.03067	.02998
1.5	.02931	.02865	.02800	.02736	.02674	.02612	.02552	.02494	.02436	.02380
1.6	.02324	.02270	.02217	.02165	.02114	.02064	.02015	.01967	.01920	.01874
1.7	.01829	.01785	.01742	.01699	.01658	.01617	.01578	.01539	.01501	.01464
1.8	.01428	.01392	.01357	.01323	.01290	.01257	.01226	.01195	.01164	.01134
1.9	.01105	.01077	.01049	.01022	$.0^{2}9957$	$.0^{2}9698$	$.0^{2}9445$	$.0^{2}9198$	$.0^{2}8957$	$.0^{2}8721$
2.0	$.0^{2}8491$	$.0^{2}8266$	$.0^{2}8046$	$.0^{2}7832$	$.0^{2}7623$	$.0^{2}7418$	$.0^{2}7219$	$.0^{2}7024$	$.0^{2}6835$	$.0^{2}6649$
2.1	$.0^{2}6468$	$.0^{2}6292$	$.0^{2}6120$	$.0^{2}5952$	$.0^{2}5788$	$.0^{2}5628$	$.0^{2}5472$	$.0^{2}5320$	$.0^{2}5172$	$.0^{2}5028$
2.2	$.0^{2}4887$	$.0^{2}4750$	$.0^{2}4616$	$.0^{2}4486$	$.0^{2}4358$	$.0^{2}4235$	$.0^{2}4114$	$.0^{2}3996$	$.0^{2}3882$	$.0^{2}3770$
2.3	$.0^{2}3662$	$.0^{2}3556$	$.0^{2}3453$	$.0^{2}3352$	$.0^{2}3255$	$.0^{2}3159$	$.0^{2}3067$	$.0^{2}2977$	$.0^{2}2889$	$.0^{2}2804$
2.4	$.0^{2}2720$	$.0^{2}2640$	$.0^{2}2561$	$.0^{2}2484$	$.0^{2}2410$	$.0^{2}2337$	$.0^{2}2267$	$.0^{2}2199$	$.0^{2}2132$	$.0^{2}2067$
2.5	$.0^{2}2004$	$.0^{2}1943$	$.0^{2}1883$	$.0^{2}1826$	$.0^{2}1769$	$.0^{2}1715$	$.0^{2}1662$	$.0^{2}1610$	$.0^{2}1560$	$.0^{2}1511$
2.6	$.0^{2}1464$	$.0^{2}1418$	$.0^{2}1373$	$.0^{2}1330$	$.0^{2}1288$	$.0^{2}1247$	$.0^{2}1207$	$.0^{2}1169$	$.0^{2}1132$	$.0^{2}1095$
2.7	$.0^{2}1060$	$.0^{2}1026$	$.0^{3}9928$	$.0^{3}9607$	$.0^{3}9295$	$.0^{3}8992$	$.0^{3}8699$	$.0^{3}8414$	$.0^{3}8138$	$.0^{3}7870$
2.8	$.0^{3}7611$	$.0^{3}7359$	$.0^{3}7115$	$.0^{3}6879$	$.0^{3}6650$	$.0^{3}6428$	$.0^{3}6213$	$.0^{3}6004$	$.0^{3}5802$	$.0^{3}5606$
2.9	$.0^{3}5417$	$.0^{3}5233$	$.0^{3}5055$	$.0^{3}4883$	$.0^{3}4716$	$.0^{3}4555$	$.0^{3}4398$	$.0^{3}4247$	$.0^{3}4101$	$.0^{3}3959$

[1] From Schlaifer (71, pp. 706–707, Table IV). By permission of McGraw-Hill Book Co., Inc.

Appendix

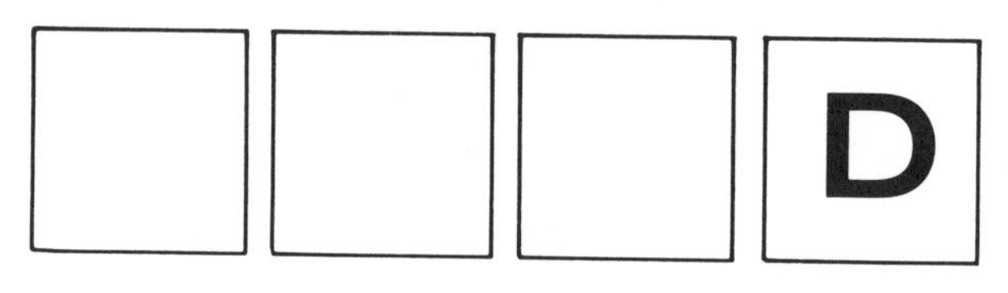

Nomograph for Computing Optimum Sample Size

This nomograph for computing optimum sample size is used to estimate a value of h, given values of D and Z, where

$$D = \frac{|\bar{x}_b - x_{\text{prior}}|}{V}$$

$$Z = \frac{V}{\sigma}\sqrt[3]{\frac{k\sigma}{c_2}}\,.$$

Using the appropriate curve for D, find the point at which this curve intersects Z. (Interpolate visually if necessary.) From this point, read horizontally across to find h. If the D curve ends so that no value of h can be determined, then no sampling is justified. Having found h, the optimum sample size is:

$$n_{\text{opt.}} = h\left(\frac{k\sigma}{c_2}\right)^{2/3}$$

The use of this nomograph is discussed in Section 5.10 and Example 5.2, pages 101–102.

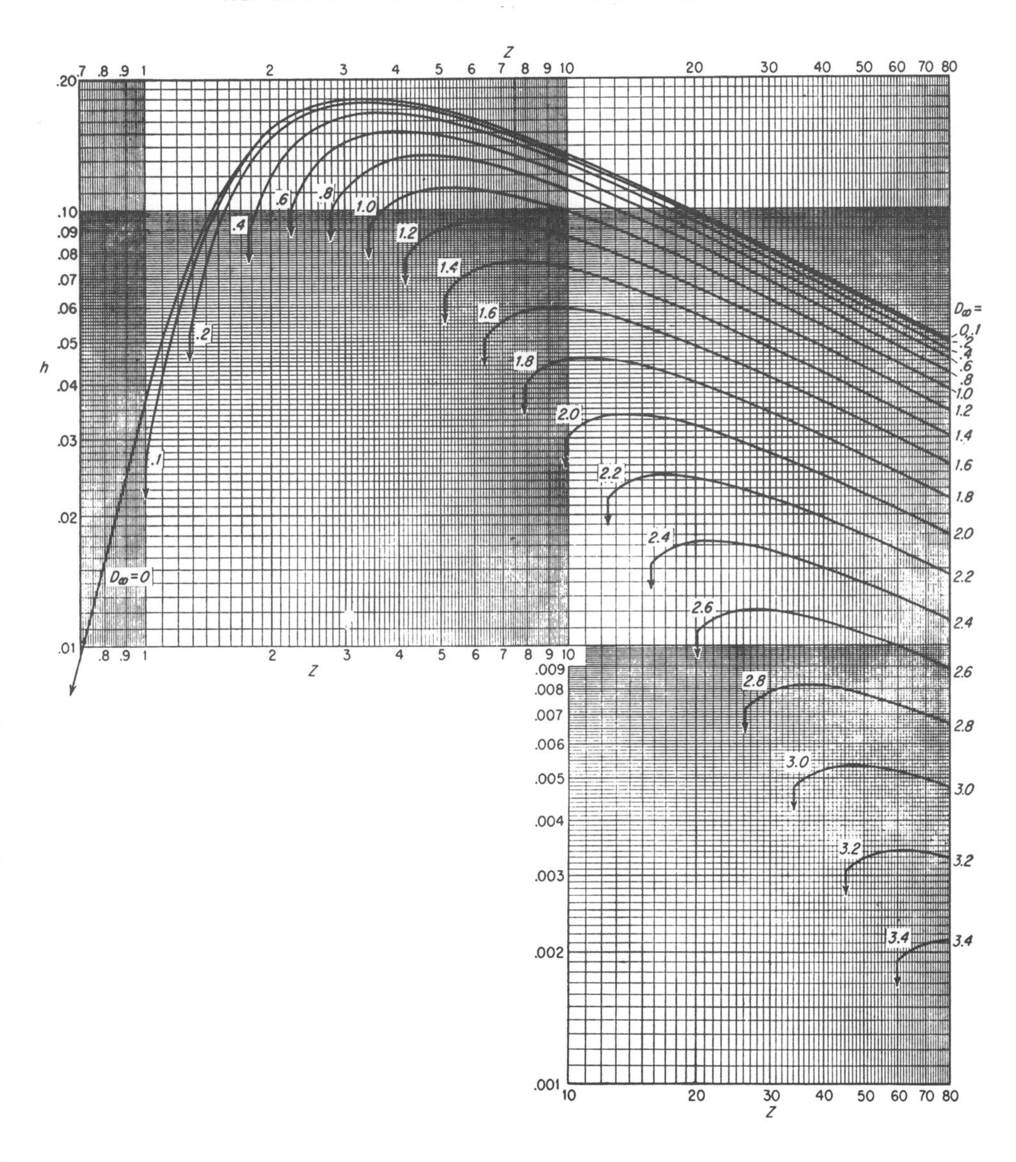

From Schlaifer (71, p. 712, Chart II). By permission of McGraw-Hill Book Co., Inc.

References

1. Aldrich, Howard E. Employment opportunities for blacks in the black ghetto: The role of white owned businesses. *American Journal of Sociology,* 1973 (May), *78,* 1403–1425.
2. Andersen, Ronald, and Anderson, Odin W. *A decade of health services: Social survey trends in use and expenditures.* Chicago: University of Chicago Press, 1967.
3. Anderson, Odin W. *Voluntary health insurance in two cities: A survey of subscriber-households.* Cambridge: Harvard University Press, 1957.
4. Anderson, Odin W., Collette, Patricia, and Feldman, Jacob J. *Changes in family medical care expenditures and voluntary health insurance: A five-year resurvey.* Cambridge: Harvard University Press, 1963.
5. Armer, Michael, and Schnaiberg, Allan. Measuring individual modernity: A near myth. *American Sociological Review,* 1972 (June), *37,* 301–316.
6. Artz, Reta, Curtis, Richard, Fairbank, Dianne, and Jackson, Elton. Community rank stratification: A factor analysis. *American Sociological Review,* 1971 (December), *36,* 985–1001.
7. Blumenthal, Monica D., Kahn, Robert L., Andrews, Frank M., and Head, Kendra B. *Justifying violence: Attitudes of American men.* Ann Arbor: Institute for Social Research, University of Michigan, 1972.
8. Bolton, Charles D. Alienation and action: A study of peace-group members. *American Journal of Sociology,* 1972 (November), *78,* 537–561.
9. Brillinger, D. R. The application of the jack-knife to the analysis of sample surveys. *Commentary* (The Journal of the British Market Research Society), 1966, *8,* 74–80.
10. Bureau of Applied Social Research. *Bibliography, from its founding to the present.* New York: BASR, Columbia University, 1967 and updates.
11. Campbell, Angus, Gurin, Gerald, and Miller, Warren E. *The voter decides.* Evanston, Ill.: Row, Peterson, 1954.

12. Campbell, Angus, and Schuman, Howard. Racial attitudes in American cities. In *Supplemental studies for the National Advisory Commission on Civil Disorders.* Washington, D.C.: U.S. Government Printing Office, 1968. (See Appendix A.)
13. Clunies-Ross, C. W. Simplifications to survey statistics. *Commentary* (The Journal of the Market Research Society), 1967, *9,* 68–76.
14. Cochran, William G. *Sampling techniques.* (2nd ed.) New York: Wiley, 1963.
15. Coleman, James S., *et al. Equality of educational opportunity.* Office of Education, U.S. Department of Health, Education, and Welfare. Washington, D.C.: U.S. Government Printing Office, 1966. 2 vols.
16. Davis, C., Hickman, J., and Novick, M. R. *A primer of decision analysis for individually prescribed instruction.* ACT Technical Bulletin No. 17. Iowa City, Ia.: American College Testing Program, 1973.
17. Davis, James A. *Great aspirations: The graduate school plans of America's college seniors.* Chicago: Aldine, 1964.
18. Deming, W. Edwards. *Sample design in business research.* New York: Wiley, 1960.
19. Dunn, Charles W., and Gove, Samuel K. Legislative reform vacuum: The Illinois case. *National Civic Review,* 1972, *61,* 441–446.
20. Edwards, Ward. Behavioral decision theory. *Annual Review of Psychology,* 1961, *12,* 473–498.
21. Edwards, Ward. The theory of decision making. *Psychological Bulletin,* 1954, *51,* 380–417.
22. Edwards, Ward, Lindman, Harold, and Savage, Leonard J. Bayesian statistical inference for psychological research. *Psychological Review,* 1963, *70,* 193–242.
23. Ericson, William A. Optimal allocation in stratified and multistage samples using prior information. *Journal of the American Statistical Association,* 1968, *63,* 964–983.
24. Ericson, William A. Optimum stratified sampling using prior information. *Journal of the American Statistical Association,* 1965, *60,* 750–771.
25. Finifter, Bernard M. The generation of confidence: Evaluating research findings by subsample replication. In Herbert L. Costner (Ed.), *Sociological methodology: 1972.* San Francisco: Jossey-Bass, 1972.
26. Flanagan, J. C., Dailey, J. T., Shaycroft, M. F., Gorham, W. A., Orr, D. B., and Goldberg, I. *Design for a study of American youth.* Boston: Houghton Mifflin, 1962.
27. Ford, W. Scott. Interracial public housing in a border city: Another look at the contact hypothesis. *American Journal of Sociology,* 1973 (May), *78,* 1426–1447.
28. Gallup, George H. (Ed.) *The Gallup Poll: Public opinion, 1935–1971.* New York: Random House, 1972. 3 vols. (Preface and introductory chapter.)
29. Garnier, Maurice. Power and ideological conformity: A case study. *American Journal of Sociology,* 1973 (September), *79,* 343–363.
30. Getman, Julius G., Goldberg, Stephen B., and Herman, Jeanne B. The National Labor Relations Board Voting Study: A preliminary report. *Journal of Legal Studies,* 1972, *1,* 233–258.
31. Glasser, Gerald J., and Metzger, Gale D. Random-digit dialing as a method of telephone sampling. *Journal of Marketing Research,* 1972, *9,* 59–64.
32, Goldman, Daniel R. Managerial mobility motivations and central life interests. *American Sociological Review,* 1973 (February), *38,* 119–125.
33. Goodman, Leo A. Snowball sampling. *Annals of Mathematical Statistics,* 1961, *32,* 148–170.
34. Greeley, Andrew, and Rossi, Peter H. *The education of Catholic Americans.* Chicago: Aldine, 1966.
35. Hansen, Morris H., Hurwitz, William N., and Madow, William G. *Sample survey methods and theory.* New York: Wiley, 1953. 2 vols.

36. Hurst, Charles E. Race, class, and consciousness. *American Sociological Review,* 1972 (December), *37,* 658–670.
37. Institute for Social Research. *List of publications.* Ann Arbor: ISR, University of Michigan.
38. Jackson, Elton F., and Curtis, Richard F. Effects of vertical mobility and status inconsistency: A body of negative evidence. *American Sociological Review,* 1972 (December), *37,* 701–713.
39. Johnstone, John W. C., and Rivera, Ramon J. *Volunteers for learning: A study of the educational pursuits of American adults.* Chicago: Aldine, 1965.
40. Kish, Leslie. *Survey sampling.* New York: Wiley, 1965.
41. Kish, Leslie, and Frankel, Martin. Balanced repeated replications for standard errors. *Journal of the American Statistical Association,* 1970, *65,* 1071–1091.
42. Kyburg, Henry E., Jr., and Smokler, Howard E. *Studies in subjective probability.* New York: Wiley, 1964.
43. Lahiri, D. B. Observations on the use of interpenetrating samples in India. *Bulletin of the International Statistical Institute,* 1957, *36* (3), 144–152.
44. Lenski, Gerhard. *The religious factor.* Garden City, N.Y.: Doubleday, 1963.
45. Lodahl, Janice B., and Gordon, Gerald. The structure of scientific fields and the functioning of university graduate departments. *American Sociological Review,* 1972 (February), *37,* 57–72.
46. Lorenz, Gerda. Aspirations of low income blacks and whites: A case of reference group processes. *American Journal of Sociology,* 1972 (September), *78,* 371–398.
47. Lowry, Ira S. (Ed.) *Housing assistance supply experiment general design report: First draft.* Santa Monica, Calif.: Rand Corporation, 1973.
48. Mahalanobis, P. C. On large-scale sample surveys. *Philosphical Transactions of the Royal Society of London, Series B,* 1946, *231,* 329–451.
49. Mahalanobis, P. C. Recent experiments in statistical sampling in the Indian Statistical Institute. *Journal of the Royal Statistical Society,* 1946, *109,* 326–370.
50. Mandell, Lewis, Katona, George, Morgan, James N., and Schmiedeskamp, Jay. *Surveys of consumers, 1971–72: Contributions to behavioral economics.* Ann Arbor: Institute for Social Research, University of Michigan, 1973.
51. McAuliffe, William E., and Gordon, Robert A. A test of Lindesmith's theory of addiction: The frequency of euphoria among long-term addicts. *American Journal of Sociology,* 1974 (January), *79,* 795–840.
52. Moser, C. A., and Kalton, G. *Survey methods in social investigation.* (2nd ed.) New York: Basic Books, 1972.
53. National Opinion Research Center. *Bibliography of publications, 1941–1960.* Chicago: NORC, 1961 and supplements.
54. National Opinion Research Center. *How to list for an area sample.* Chicago: NORC, 1964.
55. National Opinion Research Center. *NORC social research, 1941–1964: An inventory of studies and publications in social research.* Chicago: NORC, 1964.
56. Nelsen, Hart M., Yokley, Raytha, and Madron, Thomas. Ministerial roles and social actionist stance: Protestant clergy and protest in the sixties. *American Sociological Review,* 1973 (June), *38,* 375–386.
57. Neyman, Jerzy. On the two different aspects of the representative method: The method of stratified sampling and the method of purposive selection. *Journal of the Royal Statistical Society,* 1934, *97,* 558–606.
58. Novick, Melvin R. *Bayesian considerations in educational information systems.* ACT Research Report No. 38. Iowa City, Ia.: American College Testing Program, 1970.
59. Novick, Melvin R., Jackson, Paul H., Thayer, Dorothy T., and Cole, Nancy S. *Applica-*

tions of Bayesian methods to the prediction of educational performance. ACT Research Report No. 42. Iowa City, Ia.: American College Testing Program, 1971.

60. Novick, Melvin R., Lewis, Charles, and Jackson, Paul H. The estimation of proportions in *m* groups. *Psychometrika,* 1973, *38,* 19–46.
61. Orum, Anthony M., and Cohen, Roberta S. The development of political orientation among black and white children. *American Sociological Review,* 1973 (February), *38,* 62–74.
62. Ostlund, Lyman E. Interpersonal communication following McGovern's Eagleton decision. *Public Opinion Quarterly,* 1973–1974 (Winter), *37,* 601–610.
63. Perry, Paul. Election survey procedures on the Gallup Poll. *Public Opinion Quarterly,* 1960, *24,* 531–542.
64. Plackett, R. L., and Burman, P. J. The design of optimum multifactorial experiments. *Biometrika,* 1946, *33,* 305–325.
65. Politz, Alfred, and Simmons, Willard. An attempt to get the "not at homes" into the sample without callbacks. *Journal of the American Statistical Association,* 1949, *44,* 9–31.
66. Raiffa, Howard, and Schlaifer, Robert. *Applied statistical decision theory.* Boston: Harvard Business School, 1961.
67. Rand Corporation. *A million random digits with 100,000 normal deviates.* Glencoe, Ill.: Free Press, 1955.
68. Rosenthal, Robert. *Experimenter effects in behavioral research.* New York: Appleton, 1966.
69. Rossi, Peter H., Berk, Richard A., Boesel, David P., Eidson, Bettye K., and Groves, W. Eugene. Between white and black: The faces of American institutions in the ghetto. In *Supplemental studies for the National Advisory Commission on Civil Disorders.* Washington, D.C.: U.S. Government Printing Office, 1968. (See Appendix A.)
70. Savage, Leonard J. *The foundations of statistics.* New York: Wiley, 1954.
71. Schlaifer, Robert. *Probability and statistics for business decisions.* New York: McGraw-Hill, 1959.
72. Schulman, Gary I. Race, sex, and violence: A laboratory test of the sexual threat of the black male hypothesis. *American Journal of Sociology,* 1974 (March), *79,* 1260–1277.
73. Schwartz, David A. How fast does news travel? *Public Opinion Quarterly,* 1973–1974 (Winter), *37,* 625–627.
74. Selvin, Hanan C., and Stuart, Alan. Data-dredging procedures in survey analysis. *American Statistician,* 1966, *20*(3), 20–23.
75. Stephan, Frederick F., and McCarthy, Philip. *Sampling opinions.* New York: Wiley, 1958.
76. Stouffer, Samuel. *Communism, conformity, and civil liberties.* Garden City, New York: Doubleday, 1955.
77. Sudman, Seymour. On sampling of very rare human populations. *Journal of the American Statistical Association,* 1972, *67,* 335–339.
78. Sudman, Seymour. On the accuracy of recording on consumer panels: I and II. *Journal of Marketing Research,* 1964, *1,* 14–20, 69–83.
79. Sudman, Seymour. *Reducing the cost of surveys.* Chicago: Aldine, 1967.
80. Sudman, Seymour. The uses of telephone directories for survey sampling. *Journal of Marketing Research,* 1973, *10,* 204–207.
81. Sudman, Seymour, and Bradburn, Norman M. *Response effects in surveys: A review and synthesis.* Chicago: Aldine, 1974.
82. *Supplemental studies for the National Advisory Commission on Civil Disorders.* Washington, D.C.: U.S. Government Printing Office, 1968.

83. Survey Research Center. *Publications list.* Berkeley: SRC, University of California.
84. U.S. Bureau of Agricultural Economics. *Establishing a national consumer panel from a probability sample.* Marketing Research Report No. 40. Washington, D.C.: U.S. Government Printing Office, 1953.
85. U.S. Bureau of Labor Statistics. *The Consumer Price Index: History and techniques.* Bulletin 1517. Washington, D.C.: U.S. Government Printing Office, 1966.
86. U.S. Bureau of Labor Statistics. *Employment and earnings,* 1954–present, *1*–current volume. (Continuing series.)
87. U.S. Bureau of Labor Statistics. *Handbook of methods for surveys and studies.* Bulletin 1711. Washington, D.C.: U.S. Government Printing Office, 1971.
88. U.S. Bureau of Labor Statistics. *Monthly report on the labor force, 1959–1966.*
89. U.S. Bureau of the Census. *Bureau of the Census catalog.* Washington, D.C.: U.S. Government Printing Office.
90. U.S. Bureau of the Census. *Concepts and methods used in manpower statistics from the Current Population Survey.* Current Population Reports, Series P-23, No. 22. Washington, D.C.: U.S. Government Printing Office, 1967.
91. U.S. Bureau of the Census. *Current population reports.* Series P-20, P-23, P-25, P-26, P-27, P-28, P-60, and P-65. Washington, D.C.: U.S. Government Printing Office.
92. U.S. Bureau of the Census. *The Current Population Survey–A report on methodology.* Technical Paper No. 7. Washington, D.C.: U.S. Government Printing Office, 1963.
93. U.S. National Advisory Commission on Civil Disorders. *Report.* Washington, D.C.: U.S. Government Printing Office, 1968.
94. U.S. National Center for Health Statistics. *Programs and collection procedures.* Vital and Health Statistics, Series 1. Washington, D.C.: U.S. Government Printing Office, 1963–present.
95. U.S. National Center for Health Statistics. *Pseudoreplication: Further evaluation and application of the balanced half-sample technique.* Vital and Health Statistics, Series 2, No. 31. Washington, D.C.: U.S. Government Printing Office, 1969.
96. U.S. National Center for Health Statistics. *Reliability of estimates with alternative cluster sizes in the Health Interview Survey.* Vital and Health Statistics, Series 2, No. 52. Washington, D.C.: U.S. Government Printing Office, 1973.
97. U.S. National Center for Health Statistics. *Replication: An approach to the analysis of data from complex surveys.* Vital and Health Statistics, Series 2, No. 14. Washington, D.C.: U.S. Government Printing Office, 1966.
98. Wald, Abraham. *Sequential analysis.* New York: Wiley, 1947.
99. Watts, Harold W. The graduated work incentive experiments: Current progress. *American Economic Review,* 1971, *61*(2), 15–21.
100. Watts, Harold W. Graduated work incentives: An experiment in negative taxation. *American Economic Review,* 1969, *59*(2), 463–472.
101. Yancey, William L., Rigsby, Leo, and McCarthy, John D. Social position and self-evaluation: The relative importance of race. *American Journal of Sociology,* 1972 (September), *78,* 338–370.
102. Zellner, Arnold. *An introduction of Bayesian inference in econometrics.* New York: Wiley, 1971.

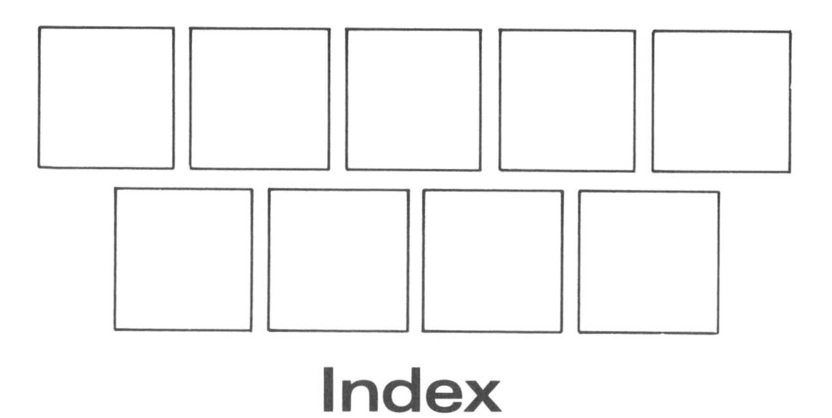

Index

C

D

I

J

K

L

M

N

W

QUANTITATIVE STUDIES IN SOCIAL RELATIONS

Consulting Editor: Peter H. Rossi

UNIVERSITY OF MASSACHUSETTS
AMHERST, MASSACHUSETTS

In Preparation

Peter Schmidt and Ann D. Witte, **THE ECONOMICS OF CRIME: *Theory, Methods, and Applications***

Peter H. Rossi, James D. Wright, and Andy B. Anderson (Eds.), **HANDBOOK OF SURVEY RESEARCH**

Published

Joan Huber and Glenna Spitze, **SEX STRATIFICATION: *Children, Housework, and Jobs***

Toby L. Parcel and Charles W. Mueller, **ASCRIPTION AND LABOR MARKETS: *Race and Sex Differences in Earnings***

Paul G. Schervish, **THE STRUCTURAL DETERMINANTS OF UNEMPLOYMENT: *Vulnerability and Power in Market Relations***

Irving Tallman, Ramona Marotz-Baden, and Pablo Pindas, **ADOLESCENT SOCIALIZATION IN CROSS-CULTURAL PERSPECTIVE: *Planning for Social Change***

Robert F. Boruch and Joe S. Cecil (Eds.), **SOLUTIONS TO ETHICAL AND LEGAL PROBLEMS IN SOCIAL RESEARCH**

J. Ronald Milavsky, Ronald C. Kessler, Horst H. Stipp, and William S. Rubens, **TELEVISION AND AGGRESSION: *A Panel Study***

Ronald S. Burt, **TOWARD A STRUCTURAL THEORY OF ACTION: *Network Models of Social Structure, Perception, and Action***

Peter H. Rossi, James D. Wright, and Eleanor Weber-Burdin, **NATURAL HAZARDS AND PUBLIC CHOICE: *The Indifferent State and Local Politics of Hazard Mitigation***

Neil Fligstein, **GOING NORTH: *Migration of Blacks and Whites from the South, 1900–1950***

Howard Schuman and Stanley Presser, **QUESTIONS AND ANSWERS IN ATTITUDE SURVEYS: *Experiments on Question Form, Wording, and Context***

QUANTITATIVE STUDIES IN SOCIAL RELATIONS

Michael E. Sobel, LIFESTYLE AND SOCIAL STRUCTURE: ***Concepts, Definitions, Analyses***

William Spangar Peirce, BUREAUCRATIC FAILURE AND PUBLIC EXPENDITURE

Bruce Jacobs, THE POLITICAL ECONOMY OF ORGANIZATIONAL CHANGE: ***Urban Institutional Response to the War on Poverty***

Ronald C. Kessler and David F. Greenberg, LINEAR PANEL ANALYSIS: ***Models of Quantitative Change***

Ivar Berg (Ed.), SOCIOLOGICAL PERSPECTIVES ON LABOR MARKETS

James Alan Fox (Ed.), METHODS IN QUANTITATIVE CRIMINOLOGY

James Alan Fox (Ed.), MODELS IN QUANTITATIVE CRIMINOLOGY

Philip K. Robins, Robert G. Spiegelman, Samuel Weiner, and Joseph G. Bell (Eds.), A GUARANTEED ANNUAL INCOME: ***Evidence from a Social Experiment***

Zev Klein and Yohanan Eshel, INTEGRATING JERUSALEM SCHOOLS

Juan E. Mezzich and Herbert Solomon, TAXONOMY AND BEHAVIORAL SCIENCE

Walter Williams, GOVERNMENT BY AGENCY: ***Lessons from the Social Program Grants-in-Aid Experience***

Peter H. Rossi, Richard A. Berk, and Kenneth J. Lenihan, MONEY, WORK, AND CRIME: ***Experimental Evidence***

Robert M. Groves and Robert L. Kahn, SURVEYS BY TELEPHONE: ***A National Comparison with Personal Interviews***

N. Krishnan Namboodiri (Ed.), SURVEY SAMPLING AND MEASUREMENT

Beverly Duncan and Otis Dudley Duncan, SEX TYPING AND SOCIAL ROLES: ***A Research Report***

Donald J. Treiman, OCCUPATIONAL PRESTIGE IN COMPARATIVE PERSPECTIVE

Samuel Leinhardt (Ed.), SOCIAL NETWORKS: ***A Developing Paradigm***

Richard A. Berk, Harold Brackman, and Selma Lesser, A MEASURE OF JUSTICE: ***An Empirical Study of Changes in the California Penal Code, 1955–1971***

Richard F. Curtis and Elton F. Jackson, INEQUALITY IN AMERICAN COMMUNITIES

Eric Hanushek and John Jackson, STATISTICAL METHODS FOR SOCIAL SCIENTISTS

Edward O. Laumann and Franz U. Pappi, NETWORKS OF COLLECTIVE ACTION: ***A Perspective on Community Influence Systems***

QUANTITATIVE STUDIES IN SOCIAL RELATIONS

Walter Williams and Richard F. Elmore, **SOCIAL PROGRAM IMPLEMENTATION**

Roland J. Liebert, **DISINTEGRATION AND POLITICAL ACTION:** ***The Changing Functions of City Governments in America***

James D. Wright, **THE DISSENT OF THE GOVERNED:** ***Alienation and Democracy in America***

Seymour Sudman, **APPLIED SAMPLING**

Michael D. Ornstein, **ENTRY INTO THE AMERICAN LABOR FORCE**

Carl A. Bennett and Arthur A. Lumsdaine (Eds.), **EVALUATION AND EXPERIMENT:** ***Some Critical Issues in Assessing Social Programs***

H. M. Blalock, A. Aganbegian, F. M. Borodkin, Raymond Boudon, and Vittorio Capecchi (Eds.), **QUANTITATIVE SOCIOLOGY:** ***International Perspectives on Mathematical and Statistical Modeling***

N. J. Demerath, III, Otto Larsen, and Karl F. Schuessler (Eds.), **SOCIAL POLICY AND SOCIOLOGY**

Henry W. Riecken and Robert F. Boruch (Eds.), **SOCIAL EXPERIMENTATION:** ***A Method for Planning and Evaluating Social Intervention***

Arthur S. Goldberger and Otis Dudley Duncan (Eds.), **STRUCTURAL EQUATION MODELS IN THE SOCIAL SCIENCES**

Robert B. Tapp, **RELIGION AMONG THE UNITARIAN UNIVERSALISTS:** ***Converts in the Stepfathers' House***

Kent S. Miller and Ralph Mason Dreger (Eds.), **COMPARATIVE STUDIES OF BLACKS AND WHITES IN THE UNITED STATES**

Douglas T. Hall and Benjamin Schneider, **ORGANIZATIONAL CLIMATES AND CAREERS:** ***The Work Lives of Priests***

Robert L. Crain and Carol S. Weisman, **DISCRIMINATION, PERSONALITY, AND ACHIEVEMENT:** ***A Survey of Northern Blacks***

Roger N. Shepard, A. Kimball Romney, and Sara Beth Nerlove (Eds.), **MULTIDIMENSIONAL SCALING:** ***Theory and Applications in the Behavioral Sciences,*** Volume I – Theory; Volume II – Applications

Peter H. Rossi and Walter Williams (Eds.), **EVALUATING SOCIAL PROGRAMS:** ***Theory, Practice, and Politics***